Corn in Clay

CORN IN CLAY

Maize Paleoethnobotany in Pre-Columbian Art

Mary W. Eubanks

University Press of Florida
GAINESVILLE/TALLAHASSEE/TAMPA/BOCA RATON
PENSACOLA/ORLANDO/MIAMI/JACKSONVILLE

04 03 02 01 00 99 6 5 4 3 2 1

Library of Congress Cataloging-in-Publication Data
Eubanks, Mary W., 1947–
Corn in clay: maize paleoethnobotany in pre-Columbian art / Mary W. Eubanks.
p. cm.
Includes bibliographical references and index.
ISBN 0-8130-1669-X (alk. paper)
1. Indian pottery—Latin America. 2. Indian art—Latin America. 3. Indians—Ethnobotany. 4. Corn in art. 5. Paleoethnobotany—Latin America. 6. Latin America—Antiquities. I. Title.
E59.P8E83 1999 98-38509
633.1'5'098—dc21

The University Press of Florida is the scholarly publishing agency for the State University System of Florida, comprising Florida A&M University, Florida Atlantic University, Florida International University, Florida State University, University of Central Florida, University of Florida, University of North Florida, University of South Florida, and University of West Florida.

University Press of Florida
15 Northwest 15th Street
Gainesville, FL 32611
http://www.upf.com

In memory of
Paul C. Mangelsdorf

CONTENTS

TABLES

FIGURES

ACKNOWLEDGMENTS

The inspiration to develop a corpus of pre-Columbian pottery with depictions of naturalistic mold-made maize was sparked by Paul C. Mangelsdorf, who laid out the issues and opportunities for such a project in "Corn in Prehistoric Art," a chapter in his book *Corn: Its Origin, Evolution and Improvement*. He wrote: "The illustrations presented here will serve to show some of the possibilities that I hope future research will pursue. Indeed it is gratifying to be able to report that at least one student of archaeology is now engaged in studies of this nature" (1974, 188). At that time, I was studying to become an archaeologist and was clueless about how many years it would take to master botanical systematics and maize genetics, in addition to archaeology and art history, before I could bring this unusual, fascinating, complex interdisciplinary study to fruition. It was my good fortune that, in spite of my naïveté, Mangelsdorf had faith in me, opened doors, and encouraged me to continue the prolonged journey down the path to learning what I needed to know—essentially the equivalent of two doctoral degrees—to achieve this goal.

This work is a testimony to the wealth of surprising and invaluable information waiting to be discovered in museum collections around the world. I am indebted to those museums where I conducted research for granting permission to examine, measure, and photograph the maize jars in their collections. I want to express a special thank you to people who were particularly helpful in this effort: Clifford Evans at the Smithsonian Institution Museum of Natural History; Ignacio Bernal at the Museo Nacional de Antropología; John Paddock and Howard Leigh at the Museo Frissell de Arte Zapoteca in Mitla, Oaxaca; Gordon Eckholm, Junius Bird, Charles Spencer, Sumru Aricanli, and Luca Casparis at the American Museum of Natural History; Oscar Córdova Hernández at the Museo Arqueológico Brüning in Lambayeque; Frank Hole at the Yale University Peabody Museum of Natural History; George Kubler and Susan B. Matheson at the Yale University Art Gallery; Dorie Reents-Budet and Amy Bryan at the Duke University Art Museum; Robert Wauchope, E. Wyllys Andrews and Kathe Trujillo at the Tulane University Middle American Research Institute; Anna C. Roosevelt at the Museum of the American Indian Heye Foundation; Lee A. Parsons at the St. Louis Museum of Fine Arts; Claudia N. Medoff at the University of Pennsylvania University Museum; Maria Gaida at the Museum für Völkerkunde in Berlin; Daniel Lévine and Valérie Layet at the Musée de l'Homme in Paris; Colin McEwan and James Hamil at the British Museum; Sergio Purrin at the Musées Royaux d'Art et d'Histoire in Brussels; Doris Kurella at the Linden Museum, Stuttgart; and Helmut Schindler at the Museum für Völkerkunde in Munich.

This interdisciplinary study was particularly challenging because it required expertise in three widely divergent disciplines: botany, archaeology, and art history. Therefore, I profoundly appreciate the experts in these different fields who read and critically evaluated drafts of the manuscript at different stages of its development. I express deep gratitude to archaeologists Donald L. Brockington, Anna C. Roosevelt, and Emeline H. Richardson, to botanists Paul C. Mangelsdorf, Major M. Goodman, and Walton C. Galinat, and to art historians Philippa D. Shaplin and Elizabeth P. Benson. I am grateful for the help of David Higdon in developing the statistical program for numerical taxonomic analysis of maize races depicted on the pottery. I thank Tom Settlemyre for his beautiful photographs of pottery in the Mexican collections, John Spencer for the maps, and Edward J. Dunn for his assistance in Peru. A special thanks to Alexander Grobman and Ricardo Sevilla for their gracious hospitality and assistance in Peru and to Linda Shulsky, with whom I stayed while working in New York.

It has been a joy and pleasure working with my editors at the University Press of Florida, Meredith Morris-Babb and Judy Goffman; with Trudie Calvert, copyeditor; and with Myra Engelhardt, art historian. A special thanks to them for their patience, support, and expert editorial guidance.

This work could never have been carried out without financial support. I thank the Research Corporation for funding the research in Mexico and the National Geographic Society for sponsoring the work in Peru and Europe. I am especially grateful to the Andrew R. Mellon Foundation and the Duke University Department of Botany for a fellowship in plant systematics that supported two years of intense, highly focused research employing the multiple approaches of molecular systematics, experimental breeding, cytogenetics, and numerical taxonomy to better understand the evolutionary history of maize and its wild relatives, during which time this book came to fruition.

1

Introduction

This archaeological and botanical study identifies corn or maize, *Zea mays* L., represented in relief on pre-Columbian urns and effigy jars. These maize depictions are impressions from molds that were formed from actual ears of maize. The ceramic facsimiles duplicate the external morphology of the botanical specimen in detail. The external characters of the ear provide the basic data for identification of the races of maize depicted on the pottery.

This unique method of realistically depicting maize occurs primarily on the pottery of two highly developed, contemporaneous pre-Columbian cultures: the Moche, also referred to as Mochica, of the North Coast of Peru, and the Zapotec of the Valley of Oaxaca in southern Mexico (see fig. 1). There are a few examples of pottery depicting mold-made maize from two later Peruvian cultures: the Chimú and the coastal Inca (Grobman et al. 1961, 110). The use of this technique to display maize realistically may have been more widespread in Mesoamerica than present evidence indicates. There have been reports of finds of mold-made mazorcas (for this and all other terms, see "Terminology" section at the end of this chapter) from an early Postclassic site in Morelos (Donovan C. Senter personal communication, November 26, 1975) and from excavations near Lake Amatitlán in highland Guatemala (Nicholas M. Hellmuth personal communication, July 8, 1974). In addition, there is at least one example of an Aztec brazier depicting mold-made maize at the Museo Nacional de Antropología in Mexico City (Eubanks 1973, 150).

Objectives

These replicas of maize provide a unique and valuable source of evidence for cultural contact between Mexico and South America. Because races of maize are distinct and tend to be local in distribution, they are potentially good indicators of trade and migration (Pickersgill 1972, 99). The value of

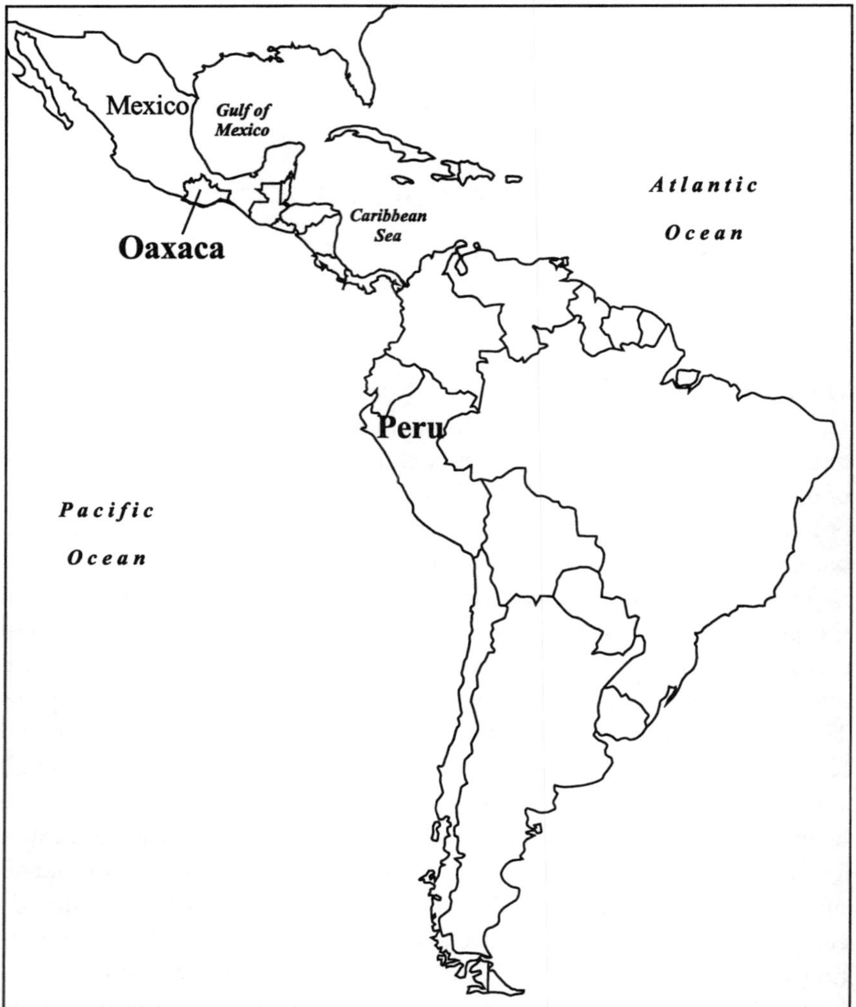

Fig. 1. Map of Mesoamerica and South America showing relative locations of Oaxaca and Peru.

geographic races as evidence for or against cultural contact, however, has been limited by the nature of archaeological materials. Finds of maize have been numerous, but they consist primarily of cobs, rarely with kernels in place or in direct association with cobs. Only occasionally have well-preserved complete maize ears have been found. They include Tularosa Cave, a dry cave site in New Mexico (Cutler 1952), and several sites on the dry, coastal desert of Peru. The Peruvian materials include some complete ears from the Central Coast site, Ancón, described by Wilhelm Reiss and Alphons Stübel (1880–87), a large collection of maize ears from the Los Cerrillos site in the Ica Valley on the South Coast (Grobman et al. 1961, 75), some ears from the Paracas Cavernas site (Yacovleff and Herrera 1934), two collections from Ocucaje Cavernas, a site contemporaneous with Paracas Cavernas (Grobman et al. 1961, 83), and approximately nineteen collections of maize from twelve archaeological sites in the area of the Upper Nazca Valley, including the Cahuachi site (Grobman et al. 1961, 83–86; Towle 1961).

Outside the above-mentioned exceptions, which come primarily from the dry South Coast of Peru, the principal maize remains at archaeological sites are cobs (Nickerson 1953, 80), and when other parts of the plant are found, they are practically never in association with cobs (Cutler 1946, 276). It has been estimated that 90 percent of maize material in archaeological collections is cobs (Lenz 1948, 353). Consequently, classification of prehistoric maize depends largely on a special set of cob characteristics. These include row number, shank diameter, cob diameter, shape of cob, width and depth of cupules, height of rachis-flaps, kernel thickness, width of lower glume, rachis diameter, and pith diameter (Nickerson 1953, 81–83). In *Zea mays,* selection has been for the ear, and its characters are central to the classification of races of maize (Mangelsdorf 1974, 102). The degree to which observable cob characters on archaeological specimens are significant for racial identifications is not known. These have not been tested to determine the relative importance of the genotype, the environment, their interaction, and sampling error on the expression of the characters (Bird 1970). Furthermore, there may be ambiguity in racial identifications of cobs as noted by Paul C. Mangelsdorf, Richard S. MacNeish, and Walton C. Galinat in reference to the races Nal Tel and Chapalote. They point out that "the two races are quite similar and it is generally not possible to distinguish the two by their cobs" (1967a, 188). The ceramic maize replicas provide corroborative evidence for identification of archaeological maize remains. They also provide evidence for a better understanding of the prehistoric distributions and relationships of races of maize and may yield information regarding the times at which certain events in the evolution of maize occurred.

If a sufficient body of botanical data indicating contact between Mexico and Peru is available, then there should be corroborative evidence from archaeology, physical anthropology, and linguistics to support the hypothesis that there was cultural contact between the two areas. A comparison of evidence supporting contact may also help to elucidate the nature of the contact situation. Important questions regarding the ceramic evidence other than the botanical identifications are: Did this apparently unique method of realistically displaying maize have a single origin or was it an independent invention? Do other aspects of the pottery reflect strong similarities, such as use, decoration, subject matter, and iconography?

In a broad perspective, the study of the ceramic evidence for prehistoric races of maize may ultimately contribute to better understanding of the development of pre-Columbian civilizations by providing data for testing the hypothesis that hybridization of maize from South America and Mesoamerica yielded more productive races better adapted to variations in climate and environmental conditions. These more productive strains, in turn, may have contributed to an expanded food supply, possibly accelerating population growth, which influenced the development of civilization during the Classic period.

The study of ceramic maize replicas makes a contribution to botany by providing new information for generating and testing hypotheses regarding the evolution of maize.

Previous Studies

J. Hébert described the use of molds by Peruvian artisans to depict fruits and vegetables in their natural forms. He reported that there were seventeen vessels depicting mold-made maize in the collection of the Musée d'Ethnographie, Paris (1902).

William Safford reported finds of funeral vases with representations of fruits and vegetables in graves near Trujillo and Chimbote on Peru's North Coast. He recognized the use of molds in the fashioning of these naturalistic representations on urns from Oaxaca, as well as on the jars from Peru's North Coast. He pointed out the botanical significance of these specimens: "Often actual specimens of maize, squashes, peanuts and fruits have been used for making molds for burial vases found interred with the mummies; and the original model has been reproduced with such accuracy that the horticultural varieties of such staples as maize and squashes are clearly discernible, and the specimens may be compared with corresponding varieties now cultivated" (1917, 14). This molding technique has also been described by Sigvald Linné (1925) and Paul Weatherwax (1954).

E. Yacovleff and F. L. Herrera (1934, 254) observed that the botanical species of the ceramic reproductions of fruits and vegetables could be identified because of the exact and precise attention to every detail. This realistic expression of fruits and vegetables was important to the Moche because these items were vital to survival and were believed to have had magical powers (Yacovleff and Herrera 1934, 252). These authors further noted that the images of maize ears were impressions from molds formed from natural models. The maize is represented with such exactitude that the perfect characteristics of the grains and ears of the varieties cultivated in different localities are distinguishable, and the excavated natural remains of these varieties are extremely rare (1934, 259).

Whereas Safford (1917) first recognized that specific varieties of fruits and vegetables were represented by these mold-made forms, Mangelsdorf and Robert G. Reeves (1938) first employed this observation in support of the theoretical supposition that Peru was the primary center of maize domestication. Bolivia is one of the few places where podcorn, their proposed ancestor of cultivated maize, is found today, but its representation by a Peruvian specimen in the Yale University Peabody Museum demonstrated that podcorn had been present in Peru in prehistoric times. In 1939, these authors illustrated several Mexican and Peruvian artifacts representing maize, including some Moche jars and an Oaxacan urn. They discussed the ceramic depictions as archaeological evidence bearing on the evolution of maize.

In their classification of Mexican maize races, E. J. Wellhausen et al. identify mold-made ceramic representations of maize on several pieces of pottery. On one Zapotec urn, they identify the ancient indigenous race Nal Tel on the basis of similarity of shape, row number, and kernel type (1952, fig. 2). This urn has also been illustrated by Eubanks (1973, plate 42).

In their classification of Peruvian races of maize, Alexander Grobman et al. (1961) identify ten "proto" races on effigy jars from museums and private collections in Peru. This was the first time identifications of ceramic specimens were supplemented by measurements of the ceramic ears. It is probable that many of the jars illustrated by Grobman et al. are also presented in this corpus, but I have been unable to verify duplication because Grobman et al. do not give the museum locations and catalog numbers of the specimens illustrated.

Mangelsdorf (1974) devoted a chapter of his book *Corn: Its Origin, Evolution and Improvement* to corn in prehistoric art. In his discussion of the importance of maize as an object of worship among cultures of the New World, he noted the depictions of maize on various pre-Columbian art forms. He emphasized the occurrence of maize impressed from molds made from actual ears on the pottery of two highly developed civilizations in Mexico and Peru and illustrated several pre-Columbian objects depicting maize.

In regard to the molding technique, Mangelsdorf (1974) posed some theoretical questions which are taken into consideration here: Did this method of depicting maize have a common origin? If it did, where did it originate? Can the time period for the diffusion and hybridization of certain races be established on the basis of the artifacts on which they are identified? He pointed out that the last question involves the problem of rumored, numerous falsifications among collections of Zapotec urns. In reference to this problem, he stated: "Indeed it is gratifying to be able to report that at least one student of archaeology is now engaged in studies of this nature" (1974, 188). In my master's thesis (Eubanks 1973), I attempted to evaluate the antiquity of twenty-six Zapotec urns depicting mold-made maize with identifiable races. The methodological approach coordinated the identifications with a rigid stylistic analysis. The urns were placed in categories of "authentic" and "suspect" according to artistic style and botanical race. When a modern race was identified, an urn's antiquity was considered "suspect." These criteria have since been proven untenable on the basis of thermoluminescence (TL) tests made by the late David W. Zimmerman, former director of the Center for Archaeometry, Washington University, St. Louis, Missouri. Zimmerman's thermoluminescence tests have been criticized because thorium levels were estimated to be higher in some of the clays than they actually are (Elizabeth P. Benson personal communication, December 31, 1982), and retesting was never done because of Zimmerman's untimely death. Since results from subsequent TL testing of Zapotec urns in the Berlin collection by a European lab are in line with Zimmerman's conclusions, however, the results of tests made on specimens in this study are believed to be reasonably accurate. Thermoluminescence involves heating a sample of ground pottery weighing approximately one gram to 400° C. The light output is measured, and the glow curve of a sample of ancient pottery is compared to the glow curve of a sample of modern pottery (Aitken 1968, 370). The tests made by Zimmerman involved measurement of the zircon particles of which Oaxa-can pottery has a high content (1971). TL tests on samples from one of the pieces labeled "suspect" showed the vessel to be at least one thousand years old (David W. Zimmerman, personal communication, February 25, 1975).

Philippa Shaplin (1975, 1978) made a stylistic analysis of a thousand photographs of Zapotec urns complemented by the results of TL tests made by Zimmerman on one hundred urns at the St. Louis Museum of Fine Art. Only 7 percent of the urns tested were proved to be fraudulent. This result led Shaplin to conclude that "a wider range of style and iconography existed in ancient Oaxaca than archaeologists have heretofore supposed" (1975, 37). Urns that have been dated by thermoluminescence and demonstrated to be of modern manufacture are illustrated in Appendix A.

Sources of Data

The corpus of photographs of vessels depicting mold-made maize replicas on which this study focuses consists of specimens in various museum collections in Peru, Mexico, Europe, and the United States. All the vessels depicting mold-made maize in relief among those collections that were available for study were examined, photographed, measured, and described. Museums contributing specimens to this corpus include the Brooklyn Museum of Art, Brooklyn, New York; Duke University Museum of Art, Durham, North Carolina; Hamburgisches Museum für Völkerkunde, Hamburg; Linden Museum, Stuttgart; Metropolitan Museum of Art, New York; Middle American Research Institute, Tulane University, New Orleans; Musée de l'Homme, Paris; Musées Royaux d'Art et d'Histoire, Brussels; Museo Nacional de Antropología y Arqueología, Lima; Museo Amano, Lima; Museo Arqueológico Brüning de Lambayeque; Museo Frissell de Arte Zapoteca, Mitla, Oaxaca; Museo Nacional de Antropología, Mexico, D.F.; Museo Rafael Larco Herrera, Lima; Museum für Völkerkunde, Berlin; Museum für Völkerkunde, Munich; Museum of the American Indian, Heye Foundation, New York; Museum Voor Land-en Völkenkunde; National Museum of Natural History, Smithsonian Institution, Washington, D.C.; St. Louis Museum of Fine Art, St. Louis; and University Museum, University of Pennsylvania, Philadelphia.

An ideal goal of the research is to establish an exhaustive corpus of all vessels depicting mold-made maize in museums around the world. Although achieving such a goal is virtually impossible and the corpus is not exhaustive, it contains a sufficient number of specimens to provide a broad selection of types of vessels, associated iconography, and races of maize that in all likelihood constitute a representative sample of pre-Columbian depictions of corn in clay. Realistic depictions of maize are extremely rare in the grand scheme of pre-Columbian art. The Museo Rafael Larco Herrera in Lima has the largest collection of Moche pottery in existence, boasting as many as 44,000 pots. This collection has, on conservative estimate, between fifty and seventy jars depicting mold-made maize in relief. Based on this comprehensive collection, depictions of realistic mold-made corn represent less than 0.2 percent of Moche pots in museum collections. The frequency of numbers of Zapotec urns with realistic maize depictions in museum collections is comparable. Although exceptional, the scientific information that can be obtained from these rare specimens can make an important contribution to our understanding of the interrelationship of human behavior with the evolution of maize, the staff of life on which agriculture and the evolution of complex social organization in the New World was founded.

Method and Theory

There are four basic assumptions in the methodology: the ceramic depictions of maize were impressed from molds; the molds were formed from real maize ears; the mold-made maize can be distinguished from stylized forms; and the race of maize can be reasonably well identified.

A maize mold from Chancay in the collection of the Museo Amano provides concrete evidence for the use of such molds in Peru (fig. 2). There is also a maize mold from Chancay in the collection of the Lowie Museum of Anthropology, University of California at Berkeley (Frank A. Norick personal communication, November 8, 1974). The existence of maize molds from Oaxaca has been confirmed by John Paddock(personal communication), who claims he has seen some.

Evidence that the molds were formed from real ears is provided by the unique characteristic of the pairing of kernel rows. The alternating rows of paired kernels often form an easily detected zigzag pattern (fig. 3). An even number of paired rows usually produces straight rows; an odd number of paired rows produces a spiral pattern (Nickerson 1953, 79). This characteristic, along with the size and shape of the ears and kernels, clearly distinguished mold-made specimens from conventionalized ones. Stylized representations of maize are depicted either with all the rows evenly aligned or with all the rows alternating. In stylizations, the kernels are often uniformly shaped, whereas in nature kernel shape varies. Figures 4 and 5 illustrate vessels with stylized depictions of maize.

Two hundred forty-nine races of maize from Latin America have been recognized and classified in eleven publications by the National Academy of Sciences–National Research Council along with the Harvard University Bussey Institute publication on the races of maize from Mexico. Racial identifications of prehistoric specimens are made, whenever possible, on the basis of comparisons of data on living races provided in the above pub-

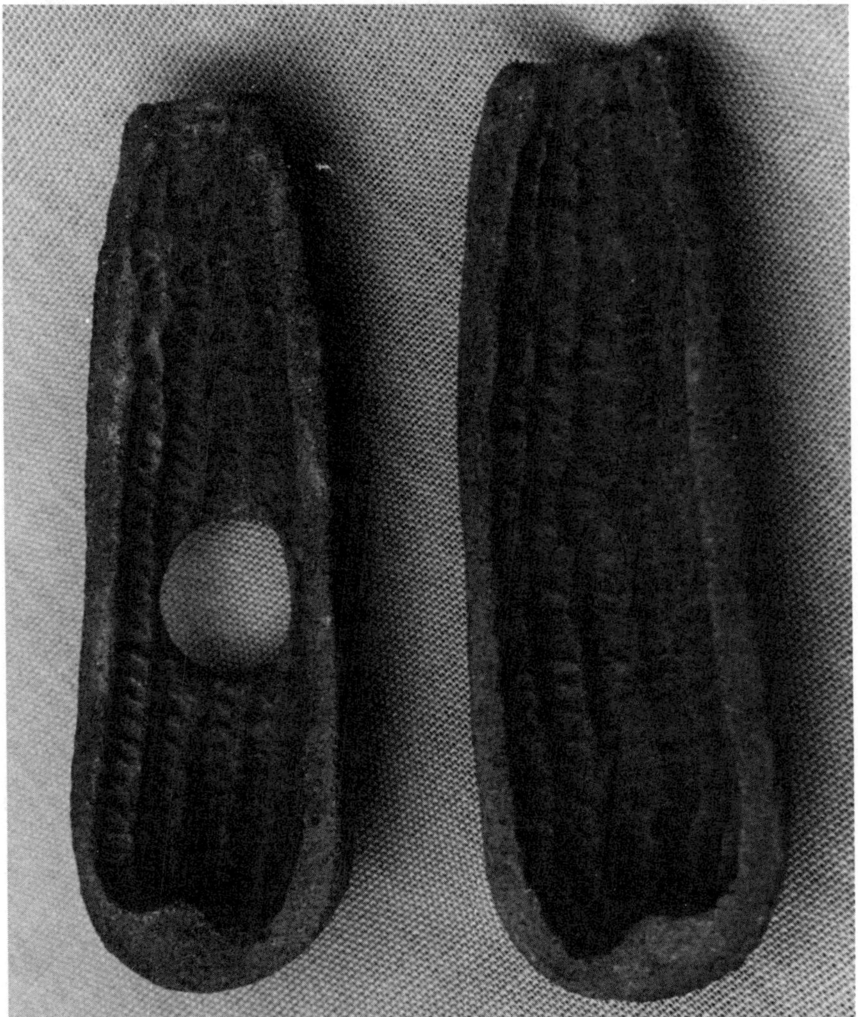

Fig. 2. Two-part maize mold, excavated in the Chancay Valley, Peru. Museo Amano 128.

lications to data from the ceramic specimens. Mangelsdorf and Gordon C. Pollard noted that "it has become a common procedure for botanists analyzing collections of prehistoric races of corn of a particular country to relate ancient specimens, so far as is possible, to the living races of that country" (1975, 49). Hugh C. Cutler recognized that the study of prehistoric maize represents a particular problem because types found "may either

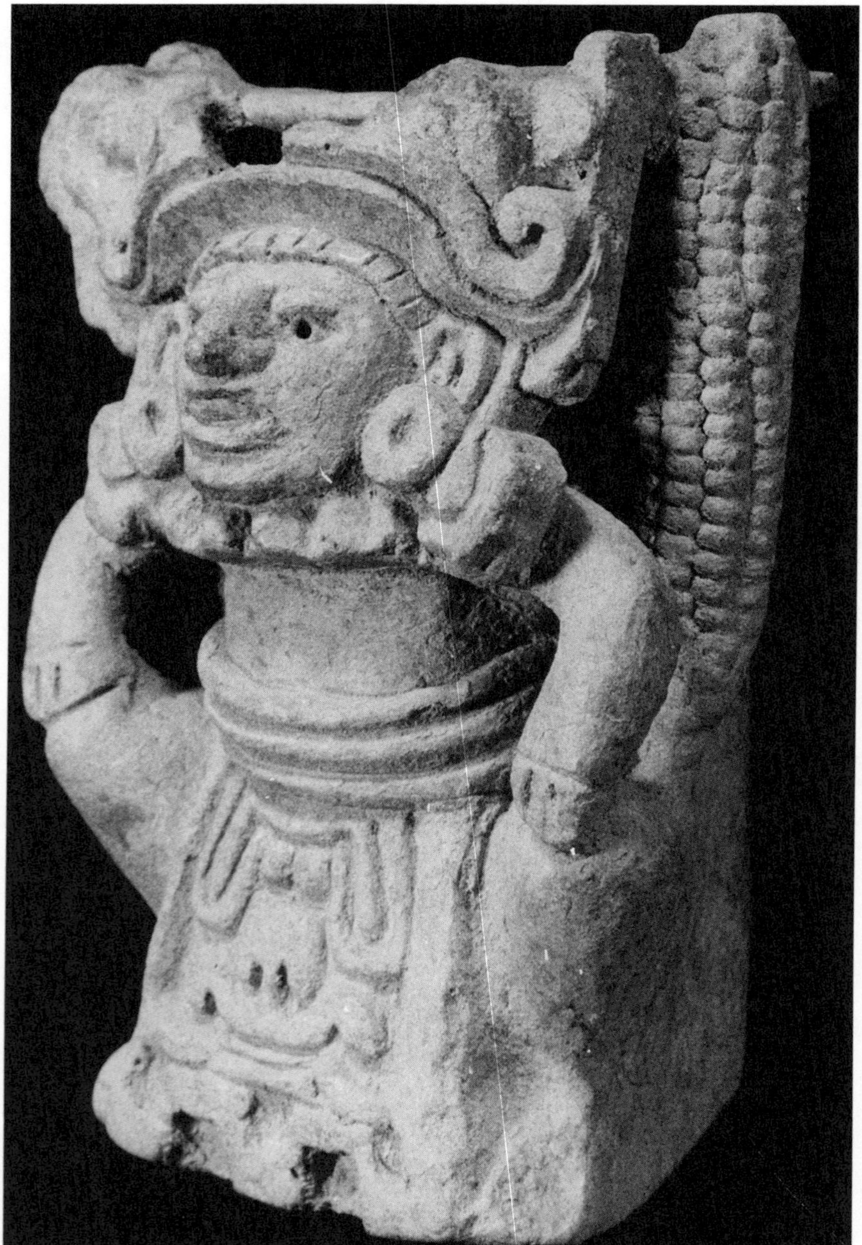

Fig. 3. Z-51. Zapotec urn. Museo Frissell de Arte Zapoteca 8269. The pairing of the kernel rows and zigzag pattern between paired rows are well illustrated by this specimen. Race of maize: Pira/Nal Tel.

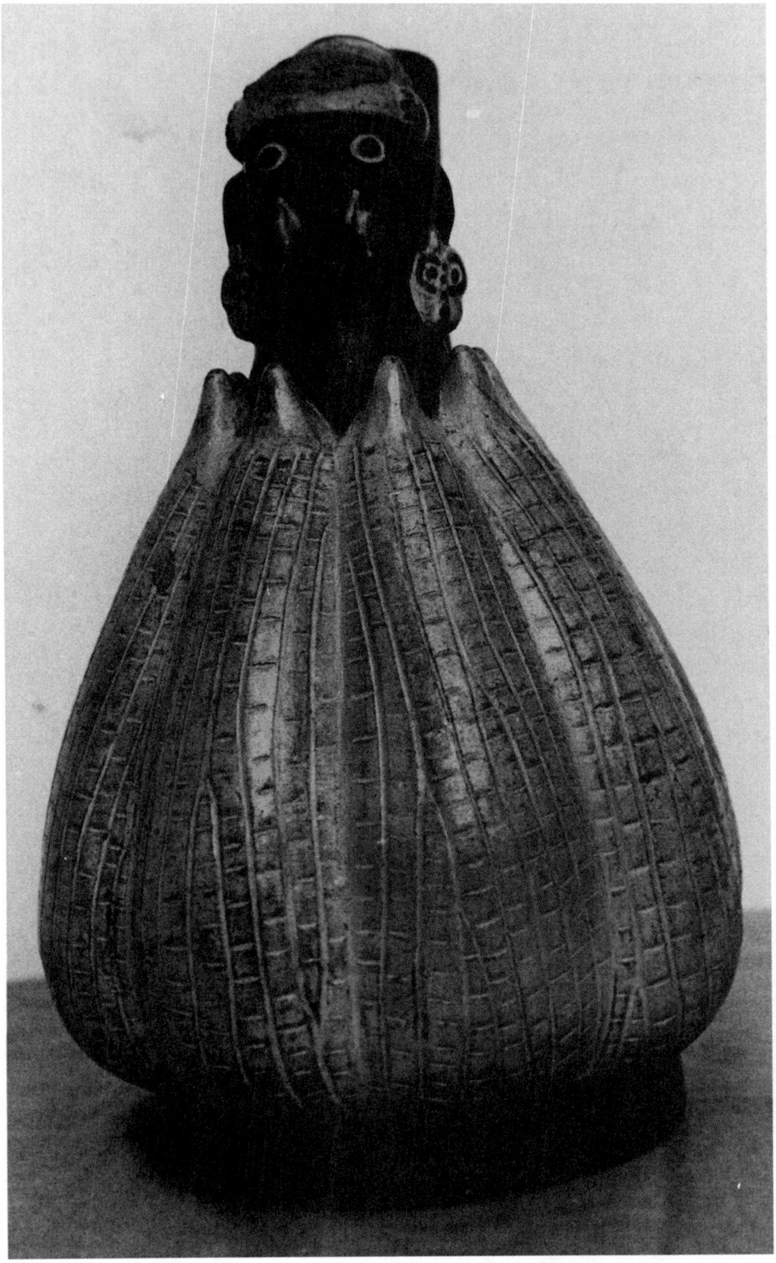

Fig. 4. Moche stirrup-spout jar. Museo Rafael Larco Herrera 2270. Stylized maize ears with incised kernels.

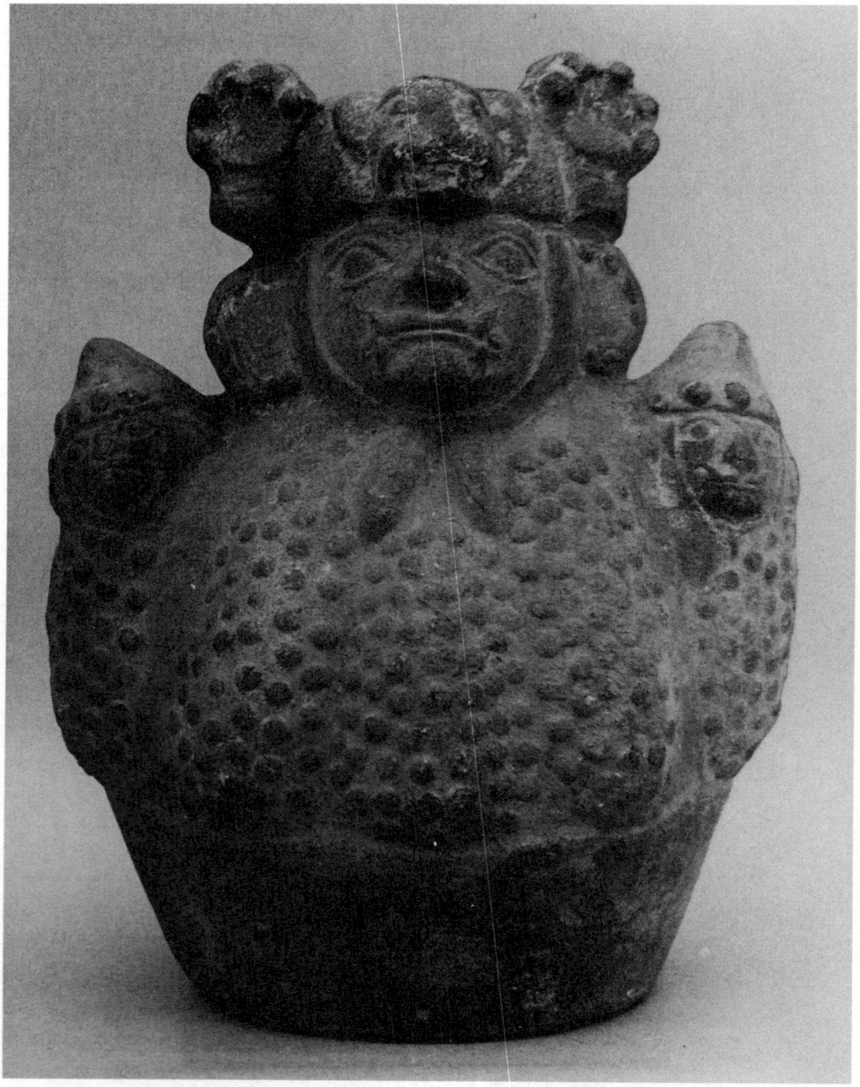

Fig. 5. Moche stirrup-spout jar. Museo Nacional de Antropología y Arqueología 42,264. The stylized ears of maize on this jar have kernels that are applied clay pellets, and every kernel row is alternating.

represent developments in the region where their remains are now found or which may in prehistoric times have been introduced from another region" (1946, 275–76). For this reason, when a race depicted on the pottery of one country could not be identified among races of that country, the identifications were sought among races in other countries.

The morphology of the ear (internal and external characters), the tassel, and the plants, as well as genetic, cytological, physiological, and agronomic characteristics, are important taxonomic considerations in the classification of maize (Wellhausen et al. 1952, 12). In *Zea mays*, however, selection has largely been for the ear. Therefore, ear characters are central to the classification of races of maize (Mangelsdorf 1974, 102). The shape and size of the ear and kernel have been stressed as particularly important diagnostic characters because they have a broad genetic base (Anderson and Cutler 1942, 72). Kernel row number, which is relatively constant within a race, is another diagnostic character (Cutler 1946, 275). These traits can be observed and recorded for the ceramic botanical facsimiles. The data provide a reliable basis for the identification of maize races depicted on pottery. All measurements are made using metal Vernier calipers, and the data are recorded as follows:

Ear shape.—Ear shape ranges from cylindrical to conical. Ears may be long-tapering or short-tapering. Some ears are short, wide to ovoid.

Ear length.—Ear length is measured from tip to base and is recorded in centimeters.

Ear diameter.—The diameter is measured at the midpoint of the length of the ear and is recorded in centimeters.

Row number.—It is assumed that half the circumference of the ear is depicted on the pottery. The count of visible rows is doubled to give the number of kernel rows for the full ear.

Kernel shape.—Kernels vary from rounded to wide and flattened in shape. They may be square-shaped or isodiametrical. Some kernels are slightly pointed or diamond-shaped. Others are so pointed that they are beaked and overlapping. These are referred to as imbricated. Some may display denting, that is, an indentation on the kernel surface. Kernel rows may be straight, spiraling, or irregularly arranged.

Kernel width.—Kernel width is obtained by measuring paired kernels, then dividing by two. The measurement is recorded in millimeters.

Kernel thickness.—The measurement for kernel thickness, recorded in millimeters, is obtained by measuring ten consecutive kernels, then dividing by ten.

Width/thickness index.—The width/thickness index is the positive number obtained by dividing kernel width by thickness. It provides an index of the kernel cross section. Cutler proposed this index because when prehistoric specimens are compared with some living races, although there may be differences in kernel sizes, there are similarities in kernel shapes (1946, 277).

Because the measured data are central to the racial identifications, it is important to know how much shrinkage may have occurred during the firing of pottery so as to gauge how closely the ceramic data approximate the data for the actual specimen from which the mold was made. After experimentation using this molding method to produce ceramic facsimiles of maize, I determined that the amount of shrinkage for the two-step process is generally less than 10 percent. Therefore, in addition to showing precise ear and kernel shape, the ceramic facsimiles provide data that closely approximate the actual measurements of the botanical specimens represented.

Although the data from the ceramic replicas may closely approximate measurements from the prehistoric specimens from which they were duplicated, I must emphasize that the data presented are measurements of maize ears that were grown more than a thousand years ago. These data are necessarily compared to data from living races. How closely the data may fit data from modern races depends on how much modern types have changed over the last thousand years. Therefore, the data cannot always be expected to fit closely. I also emphasize that although certain characters such as ear shape, kernel shape, and row number are remarkably constant within a race (Cutler 1946), there is a degree of qualitative assessment in any racial identification. Many specimens are not easily identified because they may closely approximate more than one similar, related type of maize. When such ambiguity exists, the discussion includes comparisons to reasonable alternative choices for the identification.

A basic assumption is that races of maize are distinct and tend to be local in distribution. The wide diversity of Latin American varieties of maize is attested by the fact that more than two hundred races have been described for Latin America. In addition to regional variation among races, there is also a high degree of variability within maize populations (Anderson 1946). One of the contributing factors to the variability in *Zea mays*, a cultivated plant, is that it is largely cross-fertilized and the variation is largely intraspecific (Mangelsdorf 1974, 101). Another important consideration of variability in plant populations is environmental factors (Nickerson 1953, 96). Because of the variability within maize populations, in this study of maize races, ideally several specimens are required for racial identifications. I do not ignore the problem of variability within populations, but because reliable methods for measuring ranges of variations have not been developed and tested, there are no data available with which to compare data presented here. The one exception is the study by Major M. Goodman and E. Pateriani (1969) in which they provide ranges of variation for 111 characters of fifty-five races and sub-races of maize from eastern South America. None of the races included in their sample have been identified on any specimens in this work. This problem has also been seriously addressed by Robert Bird, who studied the range of variability within maize populations through multivariate analysis of measurements of specific characters of maize collected in the Andean highlands in the departments of Huanúco and Pasco (1970).

It is also possible that measures of variability for modern maize races may not be applicable to prehistoric populations of maize. It is presumed that the Indian carefully selected for different types of maize for different purposes and carefully maintained these as pure types, just as is done today in many remote areas. Cutler observed: "Scarcely less amazing than the great diversity of maize . . . is the fact that so many varieties cultivated today have remained constant over so long a period of time. Indian fields for plant populations where optimum conditions for the evolution of new races are present. . . . That more races have not arisen is probably the result of rigid and constant selection toward definite standards. These standards vary in different districts and even today are often regulated by traditions and religious beliefs" (1946, 258–59). Edgar Anderson had presented evidence to support the assumption that variability in prehistoric maize populations was considerably less than in modern maize races. In his work in Guatemala, where there are many different varieties of maize, he was impressed by the high degree of uniformity within maize populations. He observed that in Guatemala "the plant-to-plant variation of *Zea mays* is less than any other region we have studied" (1947, 438). He suggested that one explanation for the stability of Guatemalan maize was "a rigid selection for physical type in picking seed ears. Among pure-blooded Indian farmers maize is often very carefully selected for type" (1947, 441). (I assume that Anderson's use of the term "pure-blooded" is actually an ethnic identification.)

Precise information regarding the geographical distinctiveness and the range of variability of prehistoric maize populations may ultimately be provided by a careful comparative analysis of the large numbers of maize remains in the New World archaeological record. Ninety percent of these remains consist of cobs. Therefore, such analysis awaits development of a method for distinguishing cob characters that may be reliable indicators for distinguishing maize races. In addition to a better method for analyzing archaeological maize remains, more emphasis needs to be made in archaeology on improved recovery and preservation techniques, along with careful, accurate, detailed descriptions. In his discussion of the many finds of cobs at numerous sites along the coast of Peru, Bird observed the surprising lack of literature on the remains. In reference to the published reports that do exist, he observed: "As with most reports on archaeological maize, those on Peruvian material have few data or illustrations" (1970, 123). Furthermore, archaeologists must be very careful in specifying cultural context and considering any possibility of intrusion or contamination of material (Dunn, 1978). The importance of this caveat has been dramatized by a report of the discovery of corn at the Middle Woodland Spoonville site in Michigan (Yarnell 1965, 31); it was subsequently radiocarbon dated A.D. 1950 (Munson 1966, 64).

Another problem involved in interdisciplinary botanical and archaeological research is the lack of specific interpretations regarding local developments and diffusion. Bird, a botanist, has commented on the interpretational problem posed by the "great controversy over the relative importance of local innovations and diffusion of cultural traits" (1970, 126). When considering the question of cultural diffusion, archaeologists are severely limited by the nature of materials available. They are restricted to making inferences about cultural change on the basis of archaeological remains. They rarely have access to conventional historical documentation and may have to seek documentation from legendary history. The reliability of this kind of documentation is questionable. In regard to this problem, Raymond Thompson has emphasized that "no New World archaeologist can escape the popular acceptance of the pervasive influence of the Lost Tribes of Israel" (1958, ii).

Other areas of anthropology that may provide supplementary evidence for cultural diffusion are linguistics and physical anthropology. These fields are particularly relevant to a consideration of migration. In movements of large groups of people, one might expect to find linguistic similarities, as well as resemblances of physical types.

Linguistics and physical anthropology, however, may or may not provide corroborative evidence for other kinds of cultural contact. This depends on the nature of the contact situation. Situations of culture contact may be identified archaeologically by the presence of intrusive elements from one culture into the area of another culture.

Two types of intrusive elements from cultural contact situations have been defined: site-units, a site or an occupation level in a site, which is homogeneous enough to be considered to represent the culture of a single place at a single time; and trait-units, items transported by human agency. In this book I use the trait-unit as the point of comparison. I consider the possibility that a particular trait-unit could have been a local, independent invention, even though it may occur earlier with local antecedents in another area. If this is a possibility, supplementary evidence, preferably in the form of material objects that could be transported, is necessary to establish the probability of cultural contact (Wauchope 1955, 7–8). Four criteria are used for establishing intrusion of a trait-unit: an object may be considered intrusive in an area if no local antecedent for that object is present, but there are antecedents for the object in another area where the object is also found; the two objects must be demonstrated to be contemporaneous, and the evidence of intrusion is strengthened if the object occurs at an earlier date in the area where it has antecedent developments; the possibility for physical contact between the two areas, that is, travel between the donor and recipient cultures, must be demonstrated; and the more complex and the more numerous the artifacts the stronger the case for contact between two cultures. This is a basic theoretical proposition inherent in the discussion and interpretation of the findings of this research on the ceramic evidence for the spread of prehistoric races of maize.

Terminology

There are general terms that have specific meanings in this book. There are also some special terms, the meanings of which may be unfamiliar to readers. These terms are defined under two categories: archaeological and botanical.

Archaeological Terms

Classic.—Classic refers to the civilizational development characterized by extensive construction, population growth and concentration in urban centers, class differentiation, and craft specialization. During this time, there was an active trade in raw materials and luxury items. Yet a distinctive regionalism was reflected in local art styles. According to Gordon Willey and Philip Phillips, "Technologies, elements, goods—these were exchanged; but complete idea systems remained regionalized" (1958, 187).

In Mesoamerica, the Classic period has been dated according to inscriptions on Maya monuments. Two differing correlations of dates for the

Maya monuments have received widest acceptance by archaeologists. They are the Goodman-Martinez-Thompson (GMT) correlation and the Spinden correlation. On the basis of the GMT correlation, the beginning of the Classic period is assigned to A.D. 300, and it extends to A.D. 900 (Willey 1966, 90). Adherents of the Spinden correlation place the beginning of the Classic period 260 years earlier, extending it from A.D. 40 to A.D. 640 (Coe 1966, 29). In this book, the dates for the Classic period follow the GMT correlation. It is assumed that the same chronology applies to Classic period developments in Oaxaca as it does for the Mayan area.

Although *Classic* is not the term used to denote this stage in most of the published literature on Andean archaeology, I decided to employ this term cross-culturally for the Andean region and Mesoamerica in accordance with its use by Willey and Phillips (1958) for the sake of continuity in discussion. Other terms employed in reference to this cultural, developmental stage in Peru include Early Intermediate (Rowe, 1962a), Florescent (Larco Hoyle 1966), and Regional Developmental (Lumbreras 1974). Willey places the beginning of this period in Peru at 200 B.C. and extends it to A.D. 600 (1971, 86). Here I use the term Classic period generally to refer to the span of time from ca. A.D. 300 to A.D. 900. In Peru, this covers the chronological periods on the North Coast, Moche III, IV, IVA, and possibly the beginning of Moche V.

Monte Albán I–V. Monte Albán, a mountain just outside the city of Oaxaca, was a Zapotec political and religious capital for centuries, and the chronological periods for the Valley of Oaxaca, based on changes in ceramic styles, are named for this site.

Urns with representations of mold-made maize are assignable to Monte Albán III and IV, which roughly correspond to the Classic period. Monte Albán III is subdivided into Monte Albán IIIA and IIIB. Monte Albán IIIA is distinguished from IIIB on the basis of the presence or absence of stylistic features from the great Classic period site Teotihuacán in the Valley of Mexico. Ceramic styles and shapes, along with architectural features and mural painting styles during IIIA, bear similarities to the styles from Teotihuacán (Bernal 1965, 802). Monte Albán IIIB is marked by the absence of traits from outside the Valley of Oaxaca and is characterized by cultural isolation. Monte Albán IIIB has been described by Ignacio Bernal: "In contrast with the older periods, we have found nothing at this time which comes from outside. It appears that Zapotec culture turns in on itself and becomes detached from the stream of events in Mesoamerica. I believe that this complete introversion was responsible for the marked aesthetic and technical decadence in Period IV" (1965, 805). On the basis of Shaplin's observation that it was during IIIB that ornaments on urns were "stamped out in molds," it is assumed that the urns with mold-made maize first appeared in IIIB. Shaplin's statement that "the maize figures seem to be associated with the latest material at Monte Albán and may even belong to Epoch IV after the metropolis was abandoned" lends further support to this supposition (1975, 19). The cultural manifestations for Monte Albán IIIB and IV are similar and are not easily differentiated. The two periods are distinguished by the abandonment of Monte Albán, marking the end of Monte Albán IIIB (Bernal 1965, 804).

Monte Albán IIIB and IV do not reflect a distinct chronological sequence as implied by the dates for Monte Albán III, A.D. 200–900, and Monte Albán IV, A.D. 900–1200 (Flannery et al. 1967, 447). These two periods overlap chronologically. Monte Albán IIIB represents cultural manifestations at Monte Albán up until the ceremonial abandonment of the site, and Monte Albán IV represents cultural manifestations at other sites in the Oaxaca Valley, some of which were in existence at the same time as Monte Albán IIIB. Radiocarbon dates from Lambityeco, an early but well-established Monte Albán IV site, range from ca. A.D. 650 to 750 (Rabin 1970, 15). Recent dates for Monte Albán III are A.D. 150 to 900, and for Monte Albán IV are ca. A.D. 650 to 1200 (Donald L. Brockington personal communication).

Zapotec. Zapotec is one of the Indian languages spoken by people in the Valley of Oaxaca today. It is used in reference to cultural manifestations at Monte Albán because from Monte Albán IIIA "it is possible to demonstrate that there was a continuous cultural tradition in the Valley of Oaxaca from that time until the Spanish conquest and consequently, to the present day" (Bernal 1965, 802).

Moche. Moche is the term employed in reference to the Classic period culture of the Peruvian North Coast (Bennett 1939). This is the name of one of the valleys where the culture existed and is used in accordance with the archaeological custom of naming cultures for places where they are found. Moche cultural remains are found in other nearby coastal valleys as well. Other terms in the literature that refer to Moche culture include Proto-Chimú (Uhle 1913), Early Chimú (Kutscher 1950), Mochica, a language spoken by inhabitants of the North Coast at the time of arrival of the Spanish (Schaedel 1951), and Muchic (Rowe 1948).

Moche I–V. The Moche Valley is where important remains for cultural development on the North Coast of Peru have been found, and the chronological periods, based on changes in ceramic styles, have been named for this valley. The five periods are presumed by some archaeologists to have extended from A.D. 100 to A.D. 800 (Larco Hoyle 1966; Lumbreras 1974).

Moche III is believed to have begun ca. A.D. 300 (Larco Hoyle 1966). Because of the nature of archaeology on the North Coast, where controlled excavations have been limited and for which there are few and conflicting radiocarbon dates, specific dates have not been accurately assigned to the different periods. Moche IV has been divided into two subphases: IV and IVA. The five-phase chronological sequence was first suggested by Rafael Larco Hoyle (1948). The validity of these phases and their sequence were confirmed on the basis of seriation of pottery from thirty-one grave lots from site F at Moche (Donnan 1965, 117).

Technical terms used in describing ceramics, including general art historical terms that have different shades of meaning in different contexts, and more specific, unfamiliar terms are defined for their use in this book.

Motif. Motif is used to designate a dominant idea or central theme such as the association of a particular deity with maize.

Attribute. Attribute denotes an inherent characteristic, such as an object associated with or belonging to a particular personage.

Element. Element refers to a constituent part, that is, the individual components of an attribute.

Stylized or *conventionalized*. These two terms are used interchangeably to refer to representation according to a stylistic pattern rather than according to nature.

Mold-made. Mold-made refers specifically to impressions of forms from molds that were made directly from natural objects. It distinguishes this particular technique from the term molded, which may refer to any molding process.

Aspas. Aspas refers to the eye elements that are characteristic of the Zapotec deity the God of Glyph "L."

Mazorca. Mazorca is Spanish for an ear of corn, and it is used to refer to the ceramic maize replicas.

Olla. An olla is a jar with a round bottom and vertical neck.

Florero. A florero is a high-necked jar.

Botanical Terms

Race. The definition of race herein follows Anderson and Cutler (1942, 71), who define races as "a group of related individuals with enough characteristics in common to permit their recognition as a group." This is a loose definition, which they further clarify: "From the standpoint of genetics a race is a group of individuals with a significant number of genes in common, major races having a smaller number in common than do sub-races. The degree to which a species can be divided into recognizable races and sub-races will depend upon the degree to which it has been divided into distinct intra-breeding groups with distinctive combinations of genes" (1942, 71–72).

Tripsacoid. Tripsacoid describes certain characteristics attributed to the crossing of *Zea mays* and *Tripsacum,* a related wild grass. Characteristics that may be observed on the ceramic specimens include straightening of rows, reduction of row number, and a barren cob tip (Cutler 1946, 263–64, 271).

Imbricated. Imbricated refers to the overlapping of pointed kernels.

Podcorn. In podcorn, each kernel is partially or totally enclosed by its own surrounding glumes (husks). This is usually considered to be a primitive trait.

2

Moche Effigy Jars

The North Coast of Peru: Land of the Moche

The Moche occupied the North Coast of Peru extending from the Piura Valley, near the border of Ecuador, to the Casma Valley, approximately 250 miles north of Lima (fig. 6). The long, narrow Peruvian coast, one of the world's driest deserts (Gillin 1945, 4), is composed of shifting sand dunes and rocky slopes rising abruptly out of the Pacific Ocean, where there are a few rock-bound harbors. The desert plain is crosscut by many rivers of varying capacities. Some river valleys support agriculture by means of irrigation.

The greatest development of Moche culture took place in the Moche Valley (also called the Santa Catalina Valley) and the Chicama Valley, a larger valley just to the north of Moche. Moche influence spread south to the Virú, Chao, Santa, and Nepeña Valleys and north from the heartland to the Jequetepec and Lambayeque Valleys (Moseley 1994). Evidence for Moche influence has been found as far north as the Piura Valley at Cerro Vicús and Loma Negra (Benson 1972a, 10–11).

Prehistoric Chronology

The florescence of Moche culture occurred during the Early Intermediate period, which corresponds with the Classic period in Mesoamerica, and extends from ca. 200 B.C. to A.D. 600 (Moseley 1994). The Moche ceramic style prevailed from circa 100 B.C. to A.D. 700 (Donnan 1992). Moche III, which began around A.D. 300, represents the height of development. Larco Hoyle postulates that the feline god first appeared in its human form in Moche IV (1966, 92). The feline deity in its anthropomorphic manifestation with fanged human face occurs on the majority of vessels depicting maize.

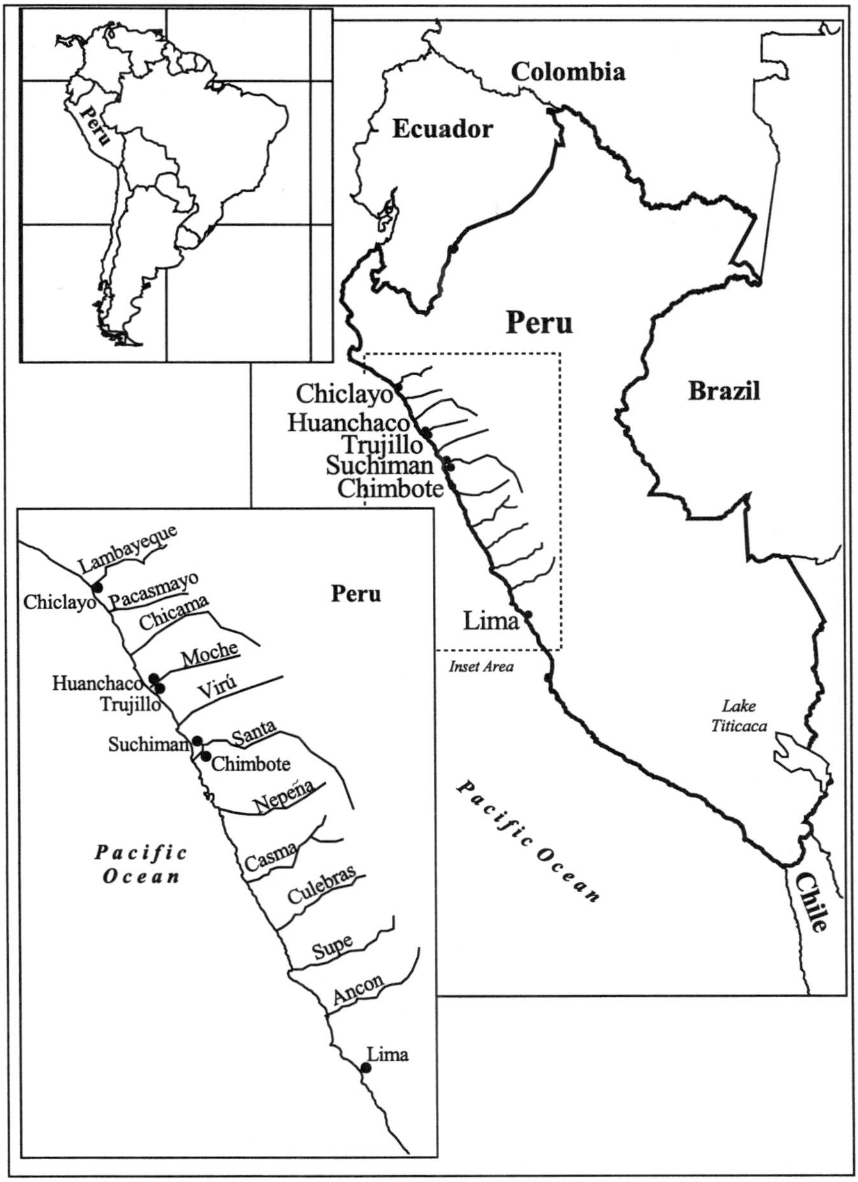

Fig. 6. Map of Peru showing North Coast proveniences of Moche jars described herein.

If Larco Hoyle's assumption is correct, then the pottery depicting mold-made maize can be assigned to Moche IV. More important, the distinctive stirrup-spout form (fig. 7) diagnostic of Moche IVA places many of the pots in Moche IV. It is impossible to assign an accurate date for the beginning of Moche IV, but it must have been around A.D. 600 and presumably continued until around the beginning of the ninth century A.D., when the more decadent artistic styles of Moche V began to appear (Lumbreras 1974). The chronology is important because if the estimated dates of ca. A.D. 600–800 for Moche IV are accepted, then Moche pottery depicting mold-made maize is contemporaneous with the occurrence of mold-made maize on Zapotec urns that can be assigned to Monte Albán IIIB-IV, ca. A.D. 600–900. Contemporaneity is an important criterion for consideration of contact between the Zapotec and Moche.

Context

Its pottery is an outstanding feature of Moche material culture. The beautiful paintings and excellent quality of the modeled forms depict all aspects of Moche life with realism unparalleled by art forms in other pre-Columbian cultures. It is a rich source of information about Moche culture, lifestyle, and worldview.

Molding was a basic technique that played a part in the manufacture of almost all ceramic objects. The majority of forms used by the Moche had local antecedents in Cupisnique, Salinar, and Gallinazo ceramics. Cupisnique, the coastal counterpart of Chavín culture which appeared in the Andean highlands ca. 800 B.C. (Lumbreras 1974, 57), appeared on the North Coast during the Middle Formative period and was centered in the Chicama Valley. It has been suggested there are direct relationships between Chavín and the Mesoamerican Formative Olmec culture (Spinden 1917). Maize agriculture first became prominent during Chavín, and a fanged feline god associated with snakes and condors appeared at this time. The stirrup-spout form and the Moche fanged deity have antecedents in the Cupisnique culture (Kan 1972). Salinar is an Upper Formative culture that followed Cupisnique in the Chicama Valley. The use of the mold can be documented by Salinar pottery. Also, designs of triangles, stepped lines, and other geometric forms painted in red and white appear on Salinar ceramics (Lumbreras 1974). Salinar cultural manifestations are followed by a new style referred to as Gallinazo. Gallinazo culture occurred in the Chicama and Virú Valleys. It disappeared with the early manifestations of Moche culture in the Chicama Valley but continued to exist in the Virú

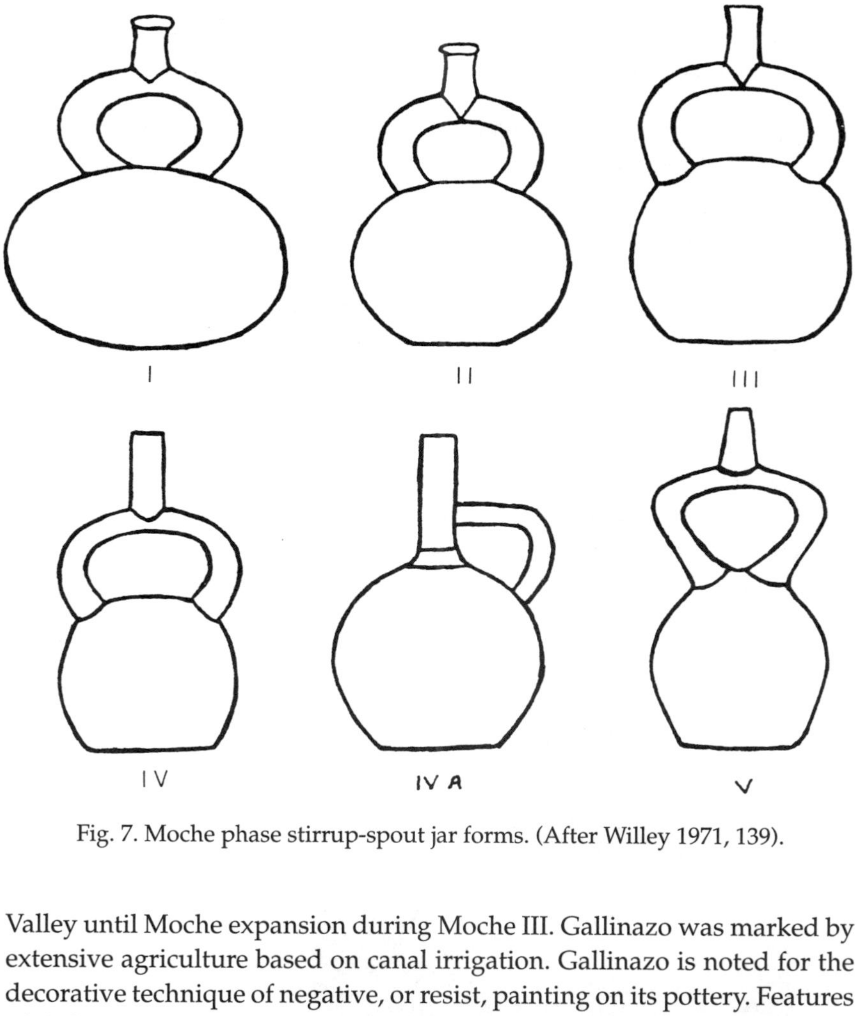

Fig. 7. Moche phase stirrup-spout jar forms. (After Willey 1971, 139).

Valley until Moche expansion during Moche III. Gallinazo was marked by extensive agriculture based on canal irrigation. Gallinazo is noted for the decorative technique of negative, or resist, painting on its pottery. Features of Gallinazo ceramics with antecedent traits in Chavín art that carry over into Moche styles are painted designs of straight and undulating lines, circles, triangles, steps, spirals, and rings painted in black on an orange surface (Strong and Evans 1952).

Molds used in the manufacture of pottery were made of fired clay formed over an actual object such as an ear of maize. When the clay dried, the object was removed and the mold fired. Positive casts were produced by pressing moist clay into the molds, and as the clay began to dry, the form was easily removed from the mold (Donnan 1965, 117–18; Donnan 1992). This technique minimizes shrinkage and is why, although a two-step process, ceramic objects made by this technique approximate the size of the objects from which they were modeled. Shrinkage is estimated to be approximately 10 percent for the entire process. The same molding technique was employed by the Zapotec. Evidence for the use of molds comes from molds excavated at archaeological sites (fig. 2), published reports of finds of molds (Uhle 1889; Kroeber 1925; Strong and Evans 1952; Thompson 1963), the visibility of the joining seam on the pottery, and repeated occurrences of identical pots (Parsons 1962).

Mold-made representations of many different species of fruits and vegetables underscore extensive diversity in Moche ethnobotany. Hermann Trimborn has observed that the plant "most commonly depicted is maize" (1968, 125). A survey that I made of depictions of fruits and vegetables illustrated in nineteen publications supports Trimborn's observation. Of eleven different plants depicted, there were thirty-three illustrations of vessels depicting maize, *Zea mays;* fourteen depicting potatoes, *Solanum* spp. (see fig. 8); seven depicting squashes, *Cucurbita maxima* and *C. moschata;* six depicting yucca, *Manihot esculenta;* five displaying peanuts, *Arachis hypogaea* (see fig. 9); three representing pacae, *Inga;* two depicting gourds,

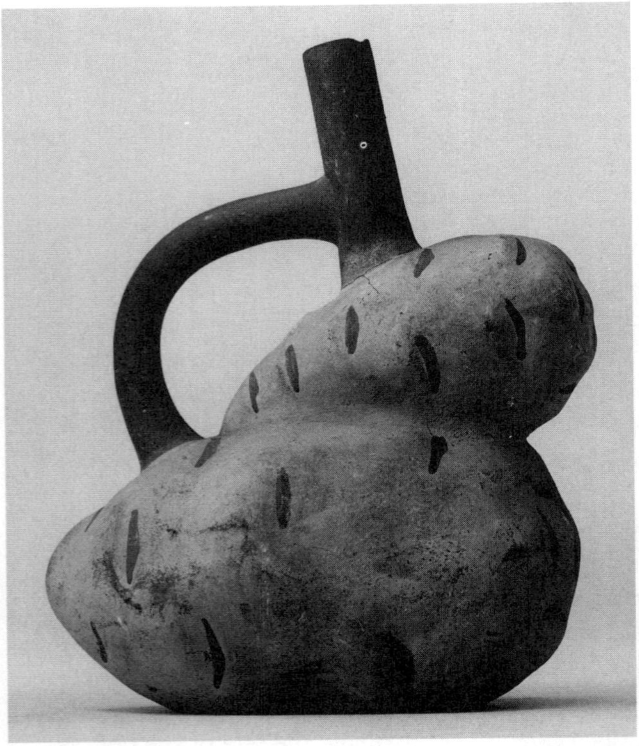

Fig. 8. Moche stirrup-spout jar in the form of a potato. Provenience: Chimbote, Peru. Catalogue number: 1188. Courtesy, Field Museum of Natural History, Chicago.

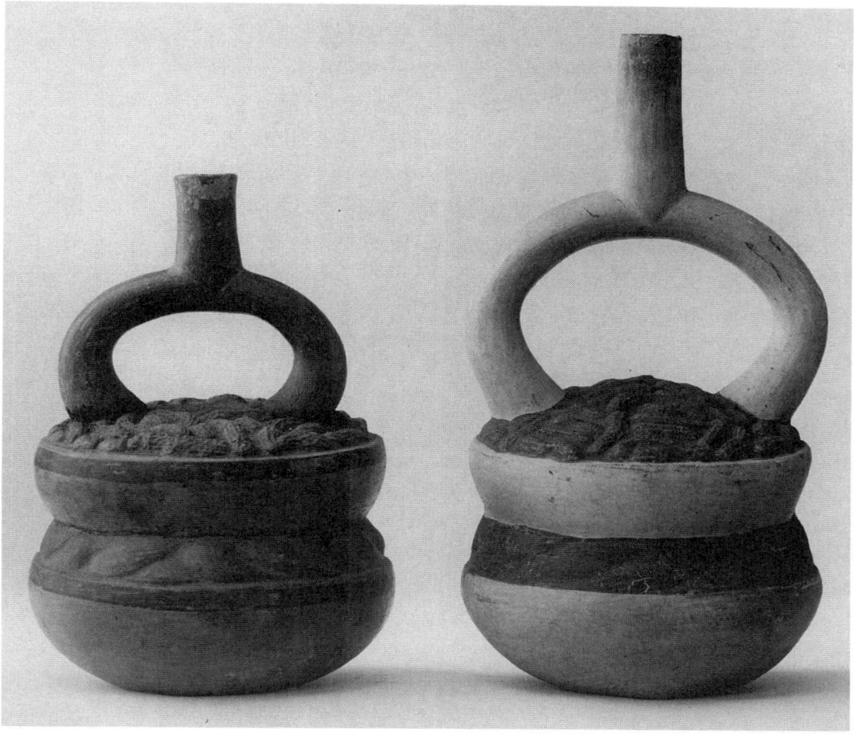

Fig. 9. Moche stirrup-spout jars depicting peanuts. Provenience: Chimbote, Peru. Catalogue numbers: 1184 and 100126. Courtesy, Field Museum of Natural History, Chicago.

Lagenaria siceraria; and one each representing papaya, *Carica candicans;* anonas or guanabana, *Anona muricata;* and achira, *Cana edulis*. Out of a total of seventy-five illustrations of fruits and vegetables in notable publications, maize represents 42 percent of the plants depicted. The plant most frequently depicted after maize is the potato. Table 2.1 provides a list of these illustrations of fruits and vegetables along with the references in which they were published. Some of the maize jars listed in table 2.1 are included in this study. Of thirty-three representations of maize, twenty-one occur on vessels in association with Ai-apaec, the Moche fanged deity. This underscores the symbolic, ritual importance of maize.

The dead were placed in rectangular graves, which were often lined with stones, cane, or adobe. Pottery jars, the primary votive offering in Moche burials, were placed around the body and in wall niches inside the graves, along with food and other offerings (Larco Hoyle 1944, 170).

Table 2.1. Moche vessels depicting plants

Plant	Association	Reference
Maize	Ai-apaec	Benson 1972b:Fig. 2-4
Maize	Ai-apaec	Fuhrman 1922:Plate 25
Maize	Ai-apaec	Weatherwax 1954:Fig. 66
Maize	Ai-apaec	Friedberg 1958:Plate IIIB
Maize	Ai-apaec	Grobman et al. 1961:Fig. 30
Maize	Ai-apaec	Grobman et al. 1961:Fig. 36
Maize	Ai-apaec	Disselhoff 1967:50
Maize	Ai-apaec	Grobman et al. 1961:Fig. 35
Maize	Ai-apaec	Friedberg 1958:Plate IIID
Maize	Ai-apaec	Grobman et al. 1961:Fig. 31
Maize	Ai-apaec	Baessler 1902-03:Fig. 290.
Maize	Ai-apaec	Schmidt 1929:Fig. 4
Maize	Ai-apaec	Weatherwax 1954:Fig. 66
Maize	Ai-apaec	Friedberg 1958:Plate IIIA
Maize	Ai-apaec	Towle 1961:Plate VII
Maize	Ai-apaec	Ville de Genève 1943:Plate 48
Maize	Ai-apaec	Grobman et al. 1961:Fig. 28
Maize	Ai-apaec	Mangelsdorf 1974:Fig. 16.14
Maize	Ai-apaec	Weatherwax 1954:Fig. 66
Maize	Ai-apaec	Baessler 1902-03:Fig. 292.
Maize	Ai-apaec	Weatherwax 1954:Fig. 66
Maize	Frog	Schmidt 1929:178
Maize	Rat	Schmidt 1929:185
Maize	Basket	Larco Hoyle 1945:Fig. 11
Maize	Bowl	Towle 1961:Plate VIIIA
Maize	Plant	Larco Hoyle 1945:12
Maize	Mound	Grobman et al. 1961:Fig. 27
Maize	Mound	Vargas 1962:Fig. 6
Maize	On a vessel	Grobman et al. 1961:Fig. 34
Maize	On a vessel	Vargas 1962:Fig. 5
Maize	Basket	Vargas 1962:Fig. 4
Potato	Prisoner	Benson 1972b:Fig. 3-22
Potato	Prisoner	Wardle 1940
Potato	Prisoner	Larco Hoyle 1945:12
Potato	Prisoner	Kosok 1965:111
Potato	Prisoner	Kosok 1965:111
Potato	Mutilated figure	Kosok 1965:111
Potato	Mutilated figure	Sawyer 1966:Fig. 77
Potato	Mutilated figures	Klein 1967:Fig. 30
Potato	Monkey	Schmidt 1929:185

(*continued*)

Plant	Association	Reference
Potato	Plain	Towle 1961:Plate XVB
Potato	Plain	Benson 1972b:Fig. 1-11
Potato	Plain	Vargas 1962:Fig. 1
Potato	Plain	Vargas 1962:Fig. 2
Potato	Plain	Vargas 1962:Fig. 3
Squash	Owl	Sawyer 1966:78
Squash	Plain	Benson 1972b:Fig. 4-7
Squash	Plain	Larco Hoyle 1938:Fig. 36
Squash	Plain	Towle 1961:Plate VIB
Squash	Plain	Towle 1961:Plate VIB
Squash	Plain	Towle 1961:Plate VIB
Squash	Plain	Vargas 1962:Fig. 12
Yucca	Ai-apaec	Benson 1972b:Fig. 5-20
Yucca	Ai-apaec	Sawyer 1966:76
Yucca	Ai-apaec	Larco Hoyle 1945:19
Yucca	Plain	Larco Hoyle 1963:Fig. 67
Yucca	Plain	Schmidt 1929:185
Yucca	Plain	Vargas 1962:Fig. 12
Peanut	Female figure	Wardle 1940
Peanut	Plain	Larco Hoyle 1945:12
Peanuts	Bowl	Towle 1961:Plate IXA
Peanuts	Bowl	Benson 1972b:Fig. 4-8
Peanuts	Bowl	Vargas 1962:Fig. 10
Pacae	Single pod	Towle 1961:Plate IXB
Pacae	Single pod	Schmidt 1929:184
Pacae	Single pod	Vargas 1962:Fig. 15
Gourd	Plain	Larco Hoyle 1963:Fig. 11
Gourd	Plain	Vargas 1962:Fig. 11
Pepino	Single fruit	Larco Hoyle 1938:Fig. 37
Pepino	Several fruits	Vargas 1962:Fig. 17
Papaya	Plain	Vargas 1962:Fig. 18
Anonas	Plain	Vargas 1962:Fig. 19
Achira	Plain	Vargas 1962:Fig. 20

Description

Most of the jars presented in this study are a beautiful oxidized orange ware. A few are highly polished black reduction ware. The orange ware vessels may be decorated with a cream-colored slip and reddish-orange or brownish-red paint. Some vessels have geometric designs painted in black, adapted from the Gallinazo style.

There are three shapes of jars: the stirrup-spout jar, a plain jar with a straight or slightly flaring neck, and an olla.

The stirrup spout is composed of two arched tubes which meet in a single cylindrical spout. The arched tubes and spout are of consistent thickness and are well rounded. The main part of the jar is spherical, and the base is flat. Some jars have an annular base. A variation of the stirrup spout is a container with a straight cylindrical spout and a curved handle attached to the spout and the jar. This form characterizes Moche IVA (fig. 7). Wendell Bennett points out that roughly 50 percent of Moche ceremonial vessels are stirrup-spout jars (1944, 101). In this corpus, 43 percent of the vessels are stirrup-spout jars.

The plain jar form is composed of a spherical body with a slightly flaring neck. These jars have a flat base. Approximately 36 percent of the jars in this corpus are of this shape.

The olla forms are flatter and wider than the other vessels. They have flat bases and short, constricted necks. Some have two small handles, one on either side of the spout. These are probably strap handles. The ears of maize on these forms are on top of the jars on the flattened area around the spout, giving the appearance of maize in a basket. They are sometimes referred to as "baskets of maize" (figs. 72–80) and are post-Moche.

The few proveniences reported for the jars were either recorded on the vessel or indicated in museum records. With the exception of the Museo Amano, none of the Peruvian museums provided card catalogs or other recorded information on the jars in their collections.

Out of a total of seventy-two Peruvian jars, forty-five represent the anthropomorphic fanged deity Ai-apaec. Ai-apaec is generally considered to be the Moche creator god whose origins can be traced back to the Chavín feline deity. The word *aiapek,* which is Yunga (Mochica) and means creator or maker (Zevallos Quiñones 1946, 178), is a later name that, according to Elizabeth P. Benson (personal communication, December 31, 1982), was "used after there were a lot of deity shifts (even within the Moche span)." There is precedent for designating the Ai-apaec figure as the major Moche deity; therefore, presentation here follows convention. Evidence that Ai-apaec was the god of agriculture comes not only from the association with maize and yucca on pottery but also from the grave of a warrior-priest in the Virú Valley excavated by William Strong and Clifford Evans. The warrior-priest was interred with, among other things, a digging stick of which the handle was carved into a representation of Ai-apaec with fangs, wearing a headdress with a feline decoration. Ai-apaec on the digging stick was accompanied by a small boy wearing a bag over his shoulder and holding three pieces of turquoise in his hand as if he were sowing seed. It has been suggested that the turquoise grains represent maize kernels and that the three snakes curling around the god's back are symbols of water and canal irrigation (Strong and Evans 1952, 199).

The most characteristic attribute of Ai-apaec is the human face with the double-fanged mouth (Lumbreras 1974). The fanged mouth occurs on all but three of the forty-one Ai-apaec figures in this corpus. The fanged mouth occurs on other creatures in pre-Columbian art and may be a sign of a supernatural creature, or it may have even broader meaning and "sometimes be a sign of sacredness or of a sacred moment—that is I think it's possible that it appears on human beings in some moment of high ritual" (Elizabeth P. Benson personal communication, December 31, 1982). Another important feature that occurs on all but three of the figures is the pointed terminus of the head. It is suggested that this cone shape may represent the tip of a maize ear. On some examples the point is completely covered with embossed kernel-like elements, while on others it is void of these kernel motifs, as if to represent an ear in which the kernels did not form over the tip, a characteristic of tripsacoid maize races.

Other Ai-apaec features include the treatment of the eyes in which the sclera are painted a cream color and the iris and pupil are unpainted. Personal adornments may include ear ornaments shaped like snake heads, bead necklaces and bracelets, and armbands. Eight figures display the characteristic elements of a Moche warrior's helmet, specifically two round "ear protectors" connected by a cord tied under the chin. These particular figures also have two round elements on the headband that are special because they are seen on only one class of Moche warrior (Elizabeth P. Benson personal communication, December 31, 1982). On ten of the jars, Ai-apaec is accompanied by two smaller figures on either side of the main figure. These small companions also have pointed heads; some have fangs, and some have an arm reaching out to touch one of the maize ears encircling the jars.

The vessels that do not represent Ai-apaec include those with rodents nibbling on an ear, a bowl containing three maize ears, several ears on a jar giving the appearance of a mound of maize, ollas representing baskets of maize, and a jar depicting a maize plant with a bird perched in the tassels. Table 2.2 summarizes the general descriptive features of the pottery.

Seventy-two Peruvian jars are presented in the following order: plain jars depicting Ai-apaec, plain jars depicting Ai-apaec accompanied by companion figures, stirrup-spout jars depicting Ai-apaec accompanied by companion figures, stirrup-spout jars depicting Ai-apaec, stirrup-spout jars

Table 2.2. Moche pottery, general features

Color		Provenience (*continued*)	
Red	7	Moche Valley	1
Red and white	42	Santa Valley	1
Red and black	2	Chiclayo	1
Red, white, and black	3	Pacasmayo	1
Black ware	15		
		Attributes of Ai-Apaec	
		Fangs	38
Shape		Old-age lines	3
Stirrup-spout	28	Companion figures	10
Moche IVA spout	4	Snake	1
Plain	24	Bare pointed head	22
Olla	15	Pointed head covered	
Bowl	1	with kernels	17
Annular base	3		
		Fillet headdress	1
Incised designs		Warrior's helmet	10
Fish	3	Snake-head ear ornaments	17
		Earspools	7
Representations		Bead necklace	8
Bowl of maize	1	Arm bands	1
Ai-apaec	41	Hair	3
Rat	3	Seated cross-legged	1
Maize plant	1	Holding ears of maize	11
Mound of maize	5		
		Painted geometric designs	
Provenience		Zigzag	3
Virú Valley	9	Stepped fret	1
Lambayeque Valley	3	Curvilinear	1
Trujillo	2	Circles	5
Huanchaco	1	Stripes	15

depicting rats, stirrup-spout jars depicting a maize plant, stirrup-spout jars of mounds of maize, a bowl of maize, and ollas resembling baskets of maize. The vessels are introduced in catalog format. Each specimen is designated by a number preceded by the letter P. After the catalog number, basic information, including the figure(s) in which the specimen is illustrated, the vessel type, provenience, museum location and museum catalog number, fabric, dimensions, and race of maize identified on the jar, is listed. Description of the specimen follows the catalog entry.

Plain Jars Depicting an Anthropomorphic Figure with Attributes of Ai-apaec

There are eighteen jars depicting an anthropomorphic figure associated with attributes of Ai-apaec. The first two illustrated here are similar (figs. 10 and 11) although not identical. Both pieces are in the collection of the Museo Nacional de Antropología y Arqueología and were excavated in the Virú Valley.

P-1. Fig. 10. Plain jar depicting an anthropomorphic figure with attributes of Ai-apaec.
Provenience: excavated in the Virú Valley.
Museo Nacional de Antropología y Arqueología KK/16583.
Fabric: red and white.
Dimensions: ht. 21.5 cm, width 13.7 cm.
Race of maize: Confite Morocho.

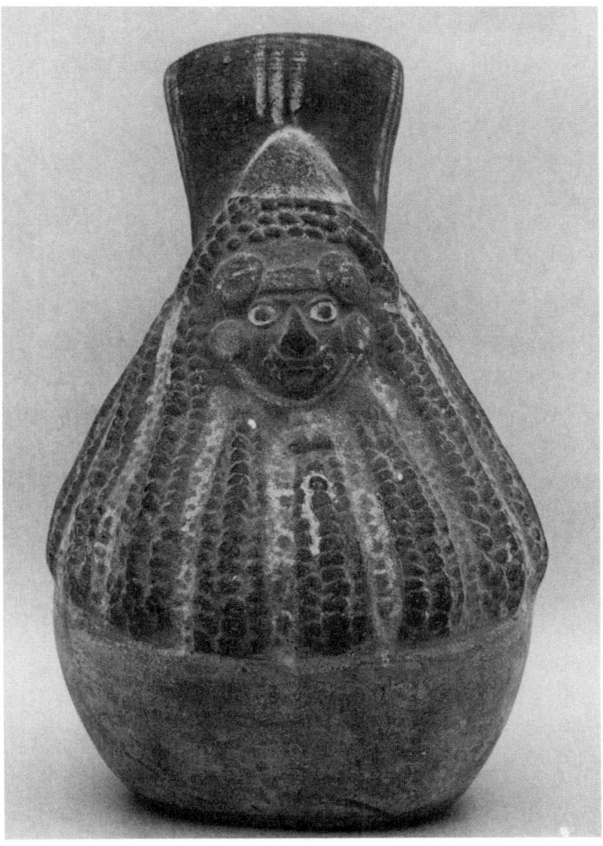

Fig. 10. P-1. Plain jar depicting an anthropomorphic figure with attributes of Ai-apaec.

The anthropomorphic figure has a pointed head with kernel elements displayed above the face and barren tip painted white. The figure wears the characteristic Moche warrior's helmet. There is a band across the forehead with concentric circle ornaments at each corner of the face. These ornaments are connected by a cord to two smaller concentric circles that appear to be earspools. The cord extends under the chin and is tied in a knot. The bottom third of the jar in unadorned. The top part of the vessel body is encircled by maize. Fourteen distinct maize ears can be counted, but the ears are blended into the jar body in such a way that a complete half round of the ear is not depicted. The plain lower third of the jar is painted white. Between the ears of maize surrounding the jar are kernels in low relief, which appear to have been produced from a mold of the same kind of maize displayed by the encircling ears. This creates the effect of the jar being continuously covered with maize, with the area in low relief contrasted with the mold-made maize because it is painted white. This gives a striped appearance to the jar, which complements the four sets of three white vertical stripes on the flaring spout of the jar.

P-2. Fig. 11. Plain jar depicting an anthropomorphic figure with attributes of Ai-apaec.
Provenience: excavated in the Virú Valley.
Museo Nacional de Antropología y Arqueología 158.
Fabric: red and white.
Dimensions: ht. 20.2 cm, width 13.2 cm.
Race of maize: Confite Morocho.

This jar is similar to the one illustrated in figure 10. The faces of the figures are identical. The maize on both jars is virtually identical; it appears to have been impressed from the same mold. This jar is slightly smaller. The main difference between the two vessels is that the first has vertical stripes on the spout, and this jar does not. Also, the lower third of the first jar is painted white. The lower third of this jar is unpainted, and a white stripe is painted around the jar at the junction of the plain section with the upper portion of the vessel.

P-3. Plain jar depicting an anthropomorphic figure with attributes of Ai-apaec.
Provenience: unknown.
Museum of the American Indian 23/196.
Fabric: red and white.
Dimensions: ht. 16.3 cm, width 13.1 cm.
Race of maize: Pira/Nal Tel.

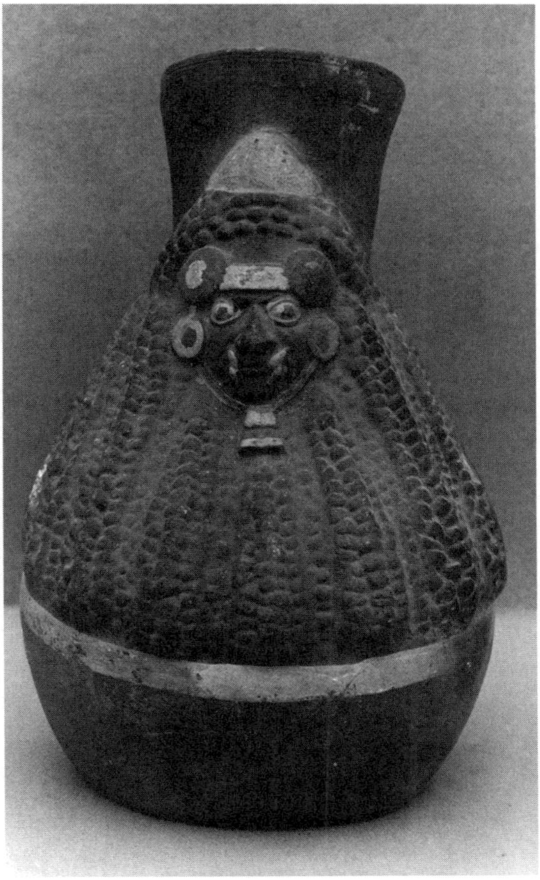

Fig. 11. P-2. Plain jar depicting an anthropomorphic figure with attributes of Ai-apaec.

This jar is similar to those described for P-1 and P-2. Four groups of three vertical parallel lines are painted on the spout, as described for P-1. A white band demarcates the area of transition from the ears of maize to the unadorned lower portion of the jar, as described for P-2. The difference is that this jar is somewhat smaller than the two preceding vessels, and the ears of maize encircling it represent a different race.

P-4. Miniature plain jar depicting an anthropomorphic figure with attributes of Ai-apaec.
Provenience: Trujillo.
Linden Museum 108636.
Fabric: red and white.
Dimensions: ht. 10.4 cm, width 5.7 cm.
Race of maize: unidentified.

This small jar is a miniature of larger jars depicting Ai-apaec seen in museum collections. It has holes on the sides of the rim for hanging. Like the previous three jars, about three-quarters of the body of the jar is covered in maize, but the ears run together and it is not possible to determine row number and length. Kernel width is 6.5 mm and thickness 3.5 mm, which is the same as Pira of Colombia. Without additional morphometric data, however, it is impossible to make a reasonable guess as to the race of maize represented. The fanged Ai-apaec figure has a pointed head with the kernel elements displayed above the face but with barren tip.

P-5. Fig. 12. Plain jar depicting an anthropomorphic figure with attributes of Ai-apaec.
Provenience: unknown.
Museo Rafael Larco Herrera J-432.
Fabric: red and white.
Race of maize: Pira/Nal Tel.

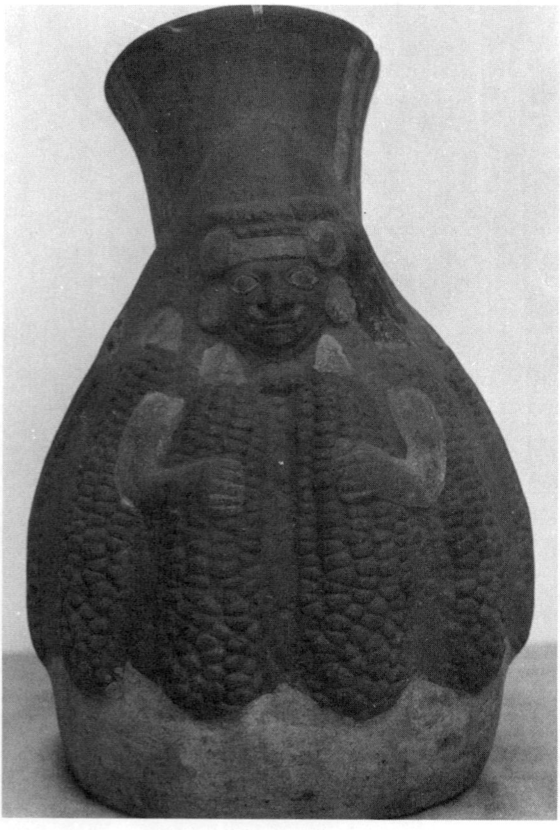

Fig. 12. P-5. Plain jar depicting an anthropomorphic figure with attributes of Ai-apaec.

This jar represents Ai-apaec with the characteristic elements of the Moche warrior's helmet. It is encircled by ten distinct ears of maize. The ears have barren tips emphasized by white paint and complemented by the bare pointed terminus of the figure's head. The figure's arms are displayed grasping two of the maize ears. The unadorned base of the jar is painted white, and there are parallel vertical white lines on the neck of the jar.

P-6. Fig. 13. Plain jar depicting an anthropomorphic figure with attributes of Ai-apaec.
Provenience: unknown.
Musée de l'Homme 78.2/752.
Fabric: red and white.
Dimensions: ht. 25.0 cm, width 15.5 cm.
Race of maize: Confite Morocho.

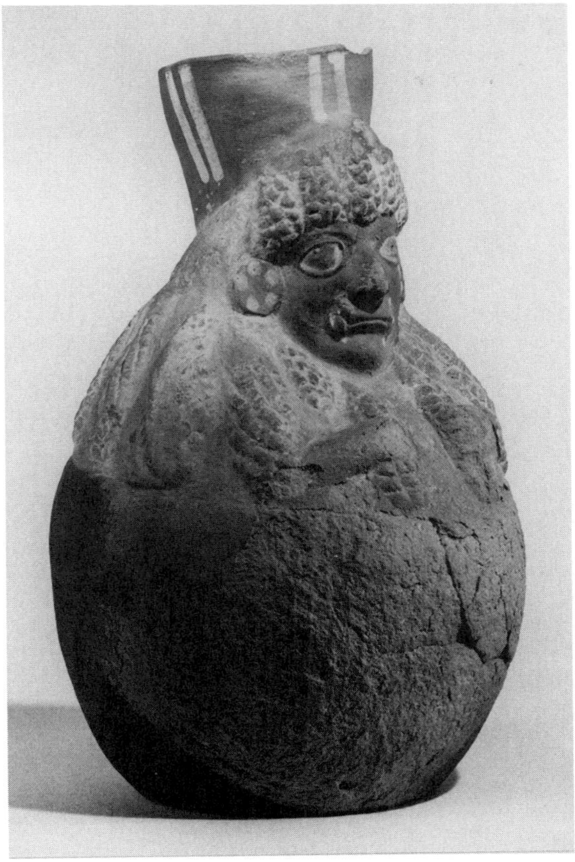

Fig. 13. P-6. Plain jar depicting an anthropomorphic figure with attributes of Ai-apaec.

This jar represents an anthropomorphic figure with some elements similar to those of jar P-5. It is encircled by twelve ears of maize in low relief. The figure's arms are displayed grasping two of the maize ears, and there are parallel vertical white lines on the neck of the jar. The figure wears earspools on which white polka dots have been painted. The fangs and sclera of the eyes are also painted white. The pointed head is covered in kernels.

P-7. Fig. 14. Plain jar depicting an anthropomorphic figure with attributes of Ai-apaec.
Provenience: unknown.
Museo Nacional de Antropología y Arqueología 1/027.
Fabric: red and white.
Dimensions: ht. 28.0 cm, width 19.5 cm.
Race of maize: Oke.

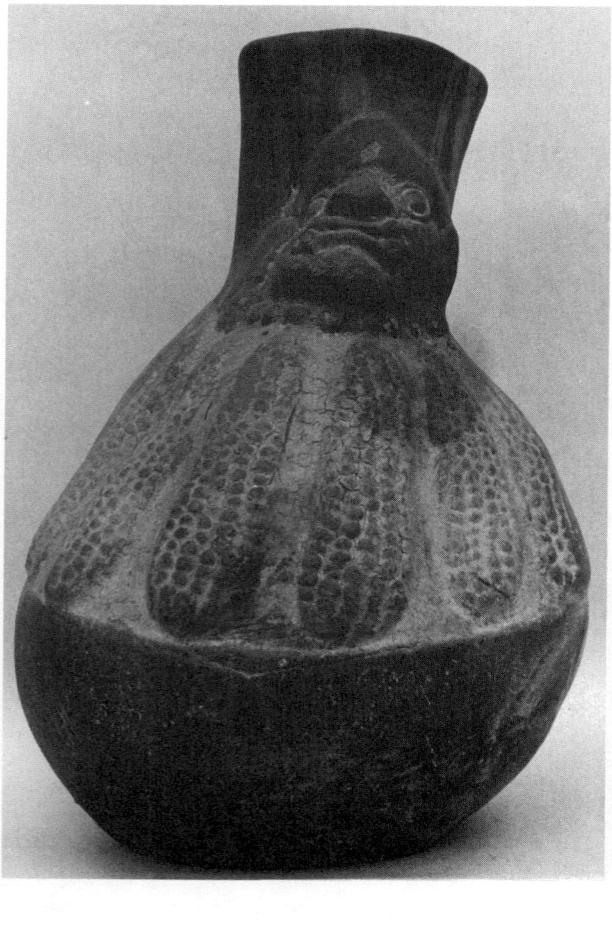

Fig. 14. P-7. Plain jar depicting an anthropomorphic figure with attributes of Ai-apaec.

This vessel has a deformed shape and thick deposits of a residue, possibly calcium carbonate, which obscures the details of the maize. The figure's head is a bare point, and its only adornment is a bead necklace. There are four groups of three parallel vertical white lines on the neck of the jar. The body of the vessel is globular. The lower part is unadorned, and the midsection is composed of eleven maize ears encircling the entire jar. Eleven ears from the same mold encircle the body of the jar.

P-8. Figs. 15 and 16. Plain jar depicting an anthropomorphic figure with attributes of Ai-apaec.
Provenience: excavated in the Lambayeque Valley.

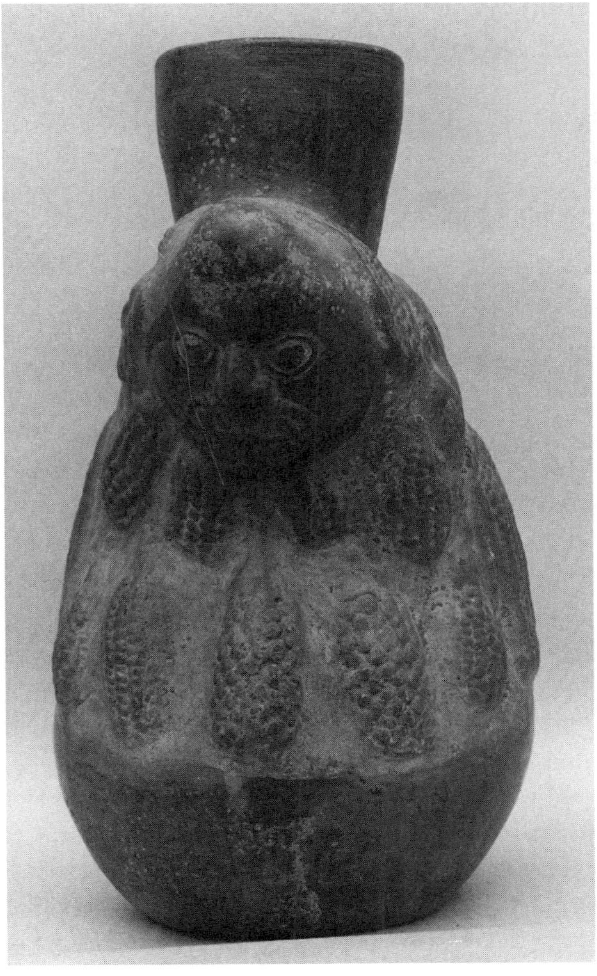

Fig. 15. P-8. Plain jar depicting an anthropomorphic figure with attributes of Ai-apaec.

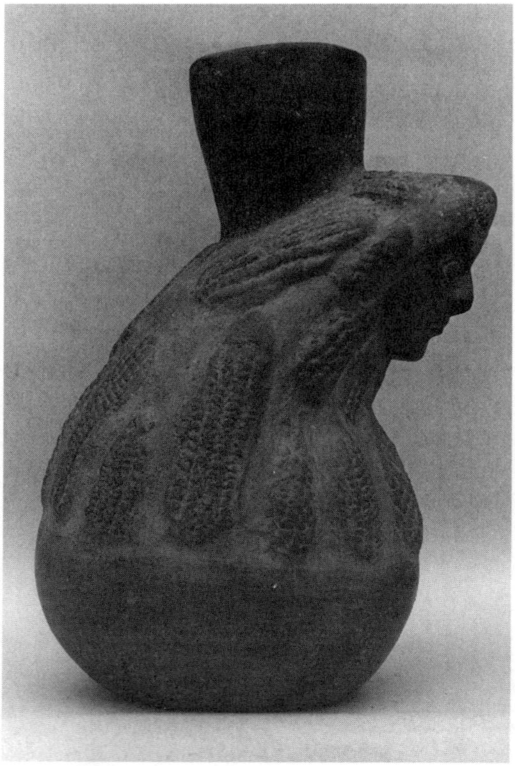

Fig. 16. P-8. Side view of fig. 15.

Museo Nacional de Antropología y Arqueología KK/16591.
Fabric: red.
Dimensions: ht. 28.0 cm, width 17.2 cm.
Race of maize: Pira/Nal Tel.

The image on this jar is stylistically different from those in the preceding figures. The mouth is much wider and is slightly open. The fangs are not painted white. The eyes are more rounded and wider. The head is tilted forward, giving the jar, from the side view (fig. 16), a shape much like that of some Moche vessels depicting gourds or squashes. The base of the jar is rounded. Its lower portion is unadorned. The midsection and the area behind the head are covered with maize ears, sporadically placed. Some of the ears are complete, whereas others display only a portion of the ear. On some of the ears, the rows are spiraling, and on some, the rows are straight. The maize replicas do not, however, appear to represent two different races; rather, they represent a race in which both types of kernel arrangements occur.

P-9. Figs. 17 and 18. Plain jar depicting an anthropomorphic figure with attributes of Ai-apaec.
Provenience: excavated in the Lambayeque Valley.
Museo Nacional de Antropología y Arqueología 2/4756.
Fabric: red.
Dimensions: ht. 23.0 cm, width 16.7 cm.
Race of maize: Pira/Nal Tel.

The fanged figure is stylistically similar to the preceding figure from the Lambayeque Valley. It has the same wide, slightly open mouth and rounded eyes. The head, however, does not tilt forward, and the neck of the jar is straight, without flaring as on other jars. The figure wears no jewelry. It has a bare, pointed head, and its ears are depicted. The base of the jar is somewhat rounded and plain, and the main body of the jar is encircled with ears of maize. Two smaller ears, one on each side of the vessel, are placed at the base of the figure's head behind its ears.

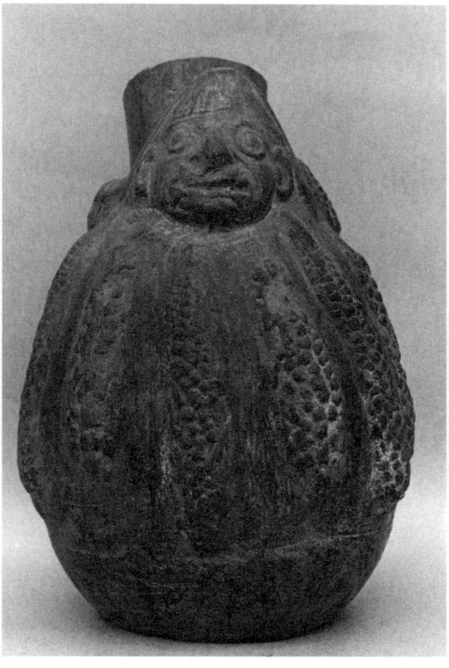

Fig. 17. P-9. Plain jar depicting an anthropomorphic figure with attributes of Ai-apaec.

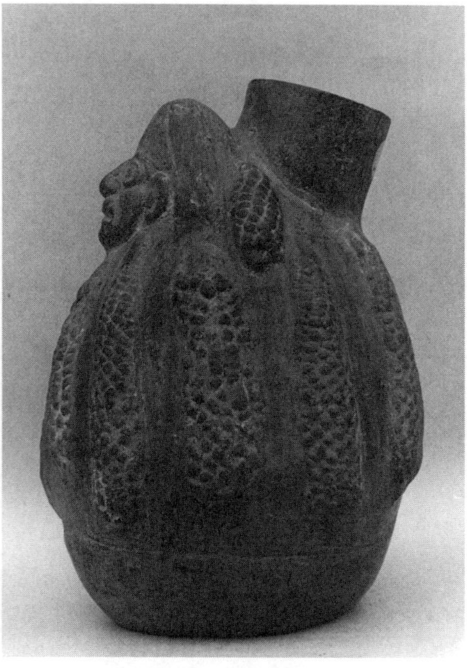

Fig. 18. P-9. Side view of fig. 17.

P-10. Fig. 19. Plain jar depicting an anthropomorphic figure with attributes of Ai-apaec.
Provenience: unknown.
Museo Rafael Larco Herrera 2295.
Fabric: red and white.
Race of maize: Confite Morocho.

The figure has large ears and wears snake-head ear ornaments. The eyes of the figure are almond-shaped, as are the eyes in figures 20–24, but its mouth is wider and slightly open, in the same style as the jars from the Lambayeque Valley. The lower third of the vessel is plain, and a white line encircling it demarcates the transition to the above section which has five maize ears on the front of the jar. The back of the jar has no maize ears on it. The head of the figure tapers into a bare pointed tip, but just below the terminus is a band of three rows of kernel-shaped elements.

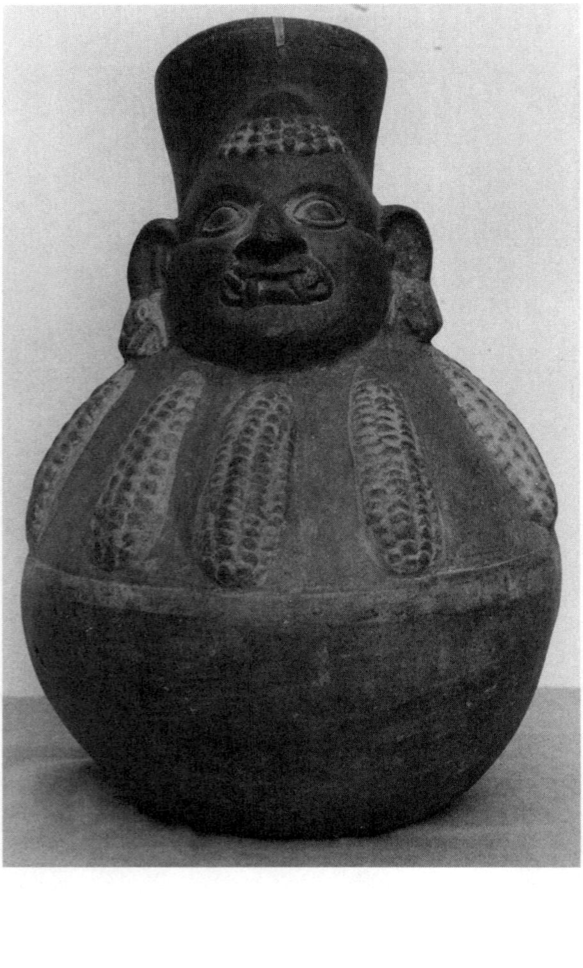

Fig. 19. P-10. Plain jar depicting an anthropomorphic figure with attributes of Ai-apaec.

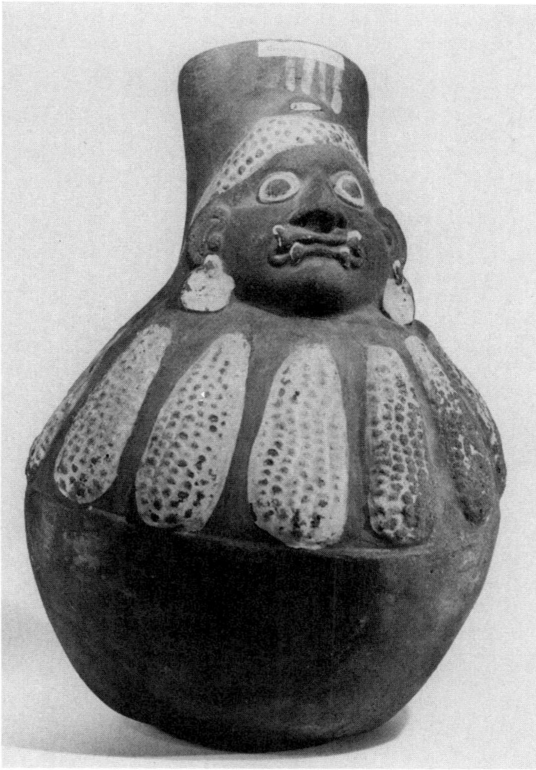

Fig. 20. P-11. Plain jar depicting an anthropomorphic figure with attributes of Ai-apaec.

P-11. Fig. 20. Plain jar depicting an anthropomorphic figure with attributes of Ai-apaec.
Provenience: unknown.
Musée de l'Homme 78.2.94.
Fabric: red and white.
Dimensions: ht. 27.5 cm, width 18.4 cm.
Race of maize: Confite Puneño/Quicheño Rojo.

The fanged figure wears snake-head ear ornaments. The lower third of the vessel is plain, and on the upper section eleven separate maize ears encircle the jar. The head of the figure forms a pointed tip covered in kernels except for a small section of the tip. The ears of maize, sclera of the eyes, fangs, earspools, and kernels of the headdress are painted white, and there are four vertical white stripes on the rim.

P-12. Figs. 21 and 22. Plain jar depicting an anthropomorphic figure with attributes of Ai-apaec.
Provenience: Chimbote.
Linden Museum 69037.

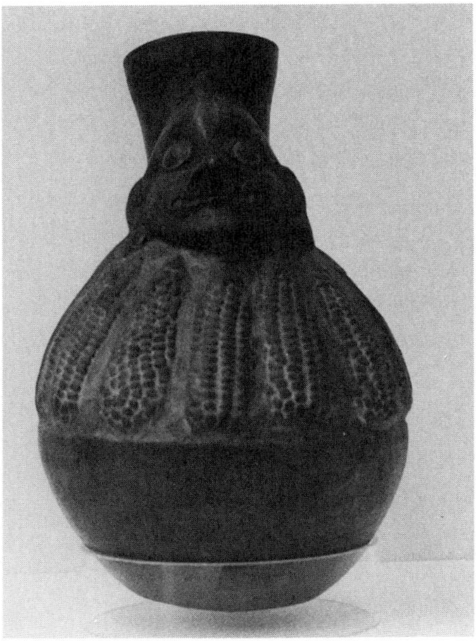

Fig. 21. P-12. Plain jar depicting an anthropomorphic figure with attributes of Ai-apaec.

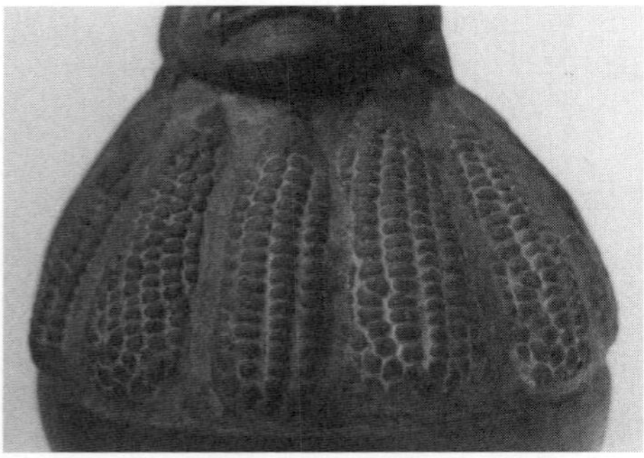

Fig. 22. P-12. Closeup of maize in fig. 21.

Fabric: red and white.
Dimensions: ht. 28.1 cm, width 18.1 cm.
Race of maize: Confite Puneño/Quicheño Rojo.

This jar depicts an anthropomorphic figure with fangs, wearing snake-head ear ornaments. Painted lines around the mouth represent old-age wrinkles. The lower section of the vessel body is plain, unpainted, and surmounted by a section of twelve ears of maize with barren tips encircling the jar. The figure's head is bare and pointed. Three different types of maize ears that are depicted on this jar are illustrated in figure 22: one with straight rows, one with slightly irregular rows at the base of the ear, and one with distinctly irregular rows.

P-13. Fig. 23. Plain jar depicting an anthropomorphic figure with attributes of Ai-apaec.
Provenience: unknown.
Museo Rafael Larco Herrera 2995.
Fabric: red and white.
Race of maize: Pira/Nal Tel.

This jar depicts Ai-apaec wearing snake-head ear ornaments and resembles jar P-12 except it does not have the characteristic fangs. The lower section of the vessel body is plain, unpainted, and surmounted by a section encircled by twelve ears of maize with barren tips. The figure's head is bare and pointed.

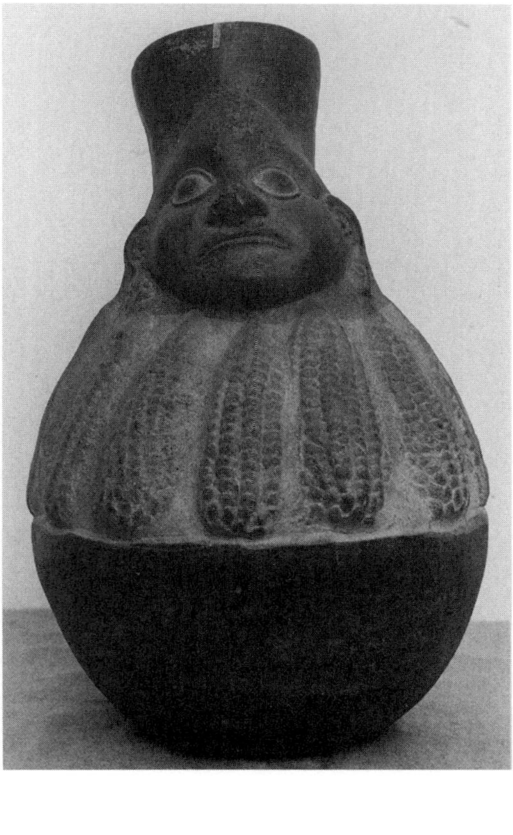

Fig. 23. P-13. Plain jar depicting an anthropomorphic figure with attributes of Ai-apaec.

P-14. Fig. 24. Plain jar depicting an anthropomorphic figure with attributes of Ai-apaec.
Provenience: unknown.
Museo Amano 59.
Fabric: red and white.
Dimensions: ht. 27.0 cm, width 18.0 cm.
Race of maize: Oke.

The figure is similar to the one described for P-13. It is unusual because it does not have the characteristic fangs associated with Ai-apaec.

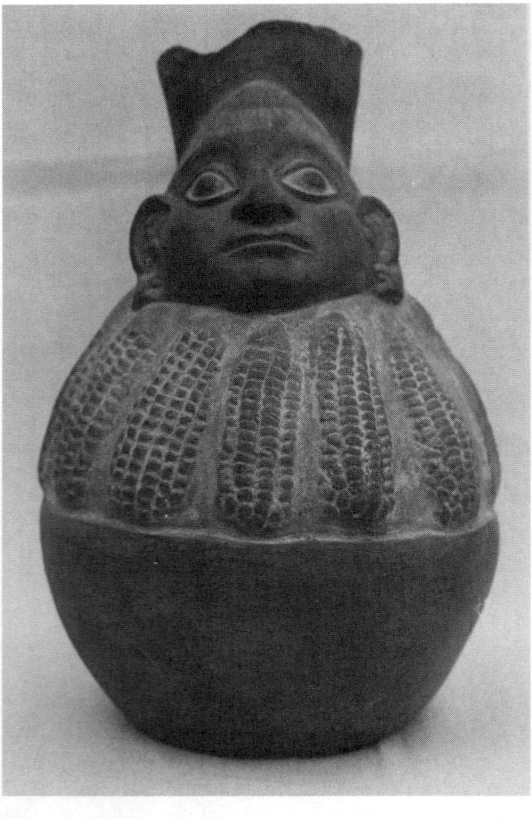

Fig. 24. P-14. Plain jar depicting an anthropomorphic figure with attributes of Ai-apaec.

P-15. Fig. 25. Plain jar depicting an anthropomorphic figure with attributes of Ai-apaec.
Provenience: Chimbote.
Museum für Völkerkunde Berlin 3727.
Fabric: red and white.
Dimensions: ht. 26.9 cm, width 16.3 cm.
Race of maize: Confite Puneño/Quicheño Rojo.

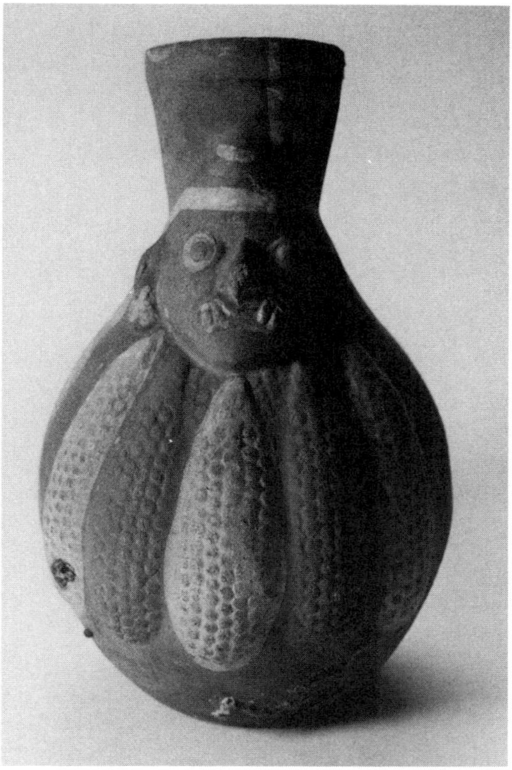

Fig. 25. P-15. Plain jar depicting an anthropomorphic figure with attributes of Ai-apaec.

This figure's head ends in a barren pointed terminus with the lower part and tip painted white. There are five ears of maize on the front of the jar. The tips of the ears are barren, and the kernels are painted white. The deity wears snake-head ear ornaments painted white. The eyes and fangs are highlighted with white paint. Instead of ears of maize on the back of the jar, kernels cover the entire area. Along the jar spout are two sets of three white vertical lines.

P-16. Fig. 26. Plain jar depicting an anthropomorphic figure with attributes of Ai-apaec.
Provenience: Chimbote.
Museum für Völkerkunde Berlin 17817.
Fabric: red and white.
Dimensions: ht. 26.9 cm, width 15.6 cm.
Race of maize: Confite Puneño/Quicheño Rojo.

This redware figure's head ends in a pointed terminus that is unadorned, and the lower part is painted white. There are five ears of maize on the front of the jar. The tips of the ears are barren, and the kernels are painted white. The deity wears snake-head ear ornaments painted white. The eyes and fangs are highlighted with white paint. Instead of ears of maize on the back of the jar, kernels cover the entire area.

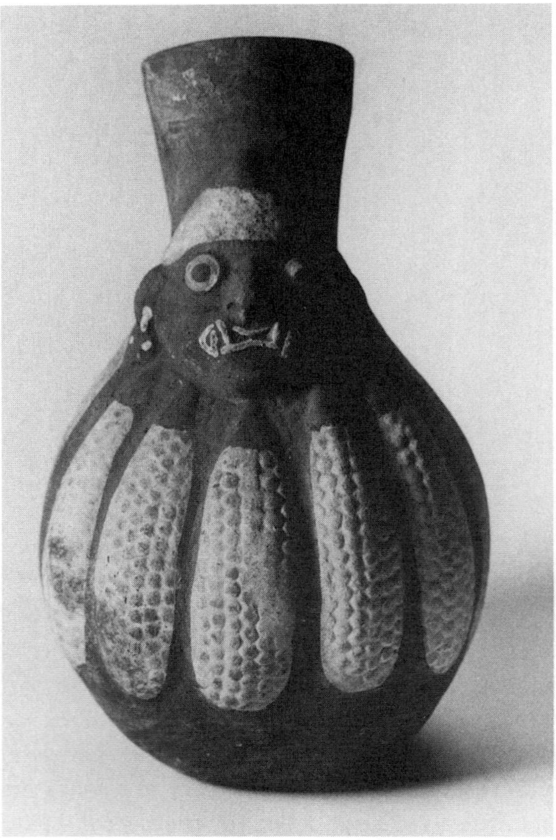

Fig. 26. P-16. Plain jar depicting an anthropomorphic figure with attributes of Ai-apaec.

P-17. Fig. 27. Plain jar depicting an anthropomorphic figure with attributes of Ai-apaec.
Provenience: Chimbote.
Museum für Völkerkunde Berlin 17819.
Fabric: red and white.
Dimensions: ht. 20.6 cm, width 13.8 cm.
Race of maize: Confite Morocho.

The redware figure's head ends in a pointed terminus that is unadorned, and the lower part is painted white. The deity wears a bead necklace and small round earspools. Six well-spaced ears of maize encircle the midsection of the jar, three in front and three in back. The lower section of the jar is

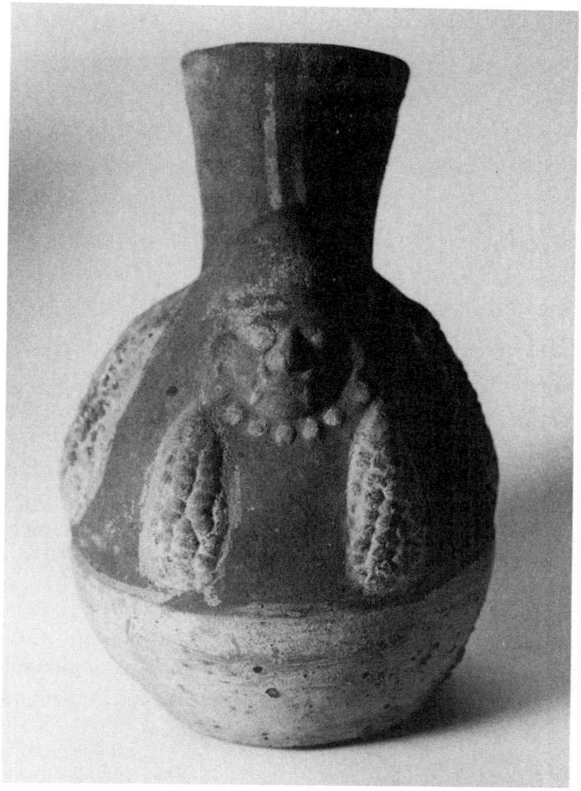

Fig. 27. P-17. Plain jar depicting an anthropomorphic figure with attributes of Aiapaec.

plain, painted white. Two vertical white lines are painted in front and back along the spout.

P-18. Fig. 28. Plain jar depicting an anthropomorphic figure with attributes of Aiapaec.
Provenience: Suchiman.
Chicago Field Museum of Natural History 4688.
Fabric: red and white.
Dimensions: ht. 31.0 cm, width 23.0 cm.
Race of maize: Confite Morocho.

The redware figure's head ends in a plain pointed terminus adorned with a white tip and white stripe. The deity wears snake-head earrings painted white. The sclera of the eyes and the fangs are white, and old-age lines are painted in dark red. Ten evenly spaced ears of maize surround the upper portion of the jar. They have barren tips that are painted white. The lower section of the jar is plain, delimited by a white stripe.

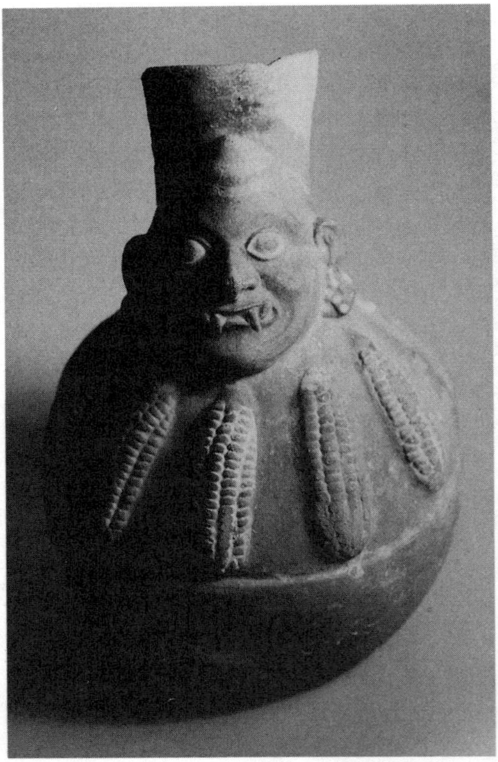

Fig. 28. P-18. Plain jar depicting an anthropomorphic figure with attributes of Ai-apaec.

P-19. Fig. 29. Plain jar depicting an anthropomorphic figure with attributes of Ai-apaec.
Provenience: unknown.
American Museum of Natural History B/8863.
Fabric: red and white.
Dimensions: ht. 23.0 cm, width 15.0 cm.
Period: Moche IV.
Race of maize: Pira/Nal Tel.

The head of this figure ends in a pointed terminus that is adorned with kernel elements and a barren tip. It has the rounded elements above the temples of the head and the tie beneath the chin that are features of a warrior's helmet. It has round earspools. The midsection of the jar is completely covered in mold-made maize in low relief. The lower section and the spout are plain.

P-20. Fig. 30. Plain jar depicting an anthropomorphic figure with attributes of Ai-apaec.
Provenience: unknown.
American Museum of Natural History 41.2/7427.
Fabric: red and white.
Dimensions: ht. 21.3 cm, width 12.5 cm.
Race of maize: Pira/Nal Tel.

The head of this figure ends in a pointed terminus that is a barren tip. It has the rounded elements above the temples of the head and the tie beneath the chin that are features of a warrior's helmet. It has round earspools. The midsection of the jar is completely covered in mold-made maize in low relief. The lower section and the spout are plain. The maize, helmet, fangs, and sclera of the eyes are painted white, and there are double white vertical parallel lines on the spout.

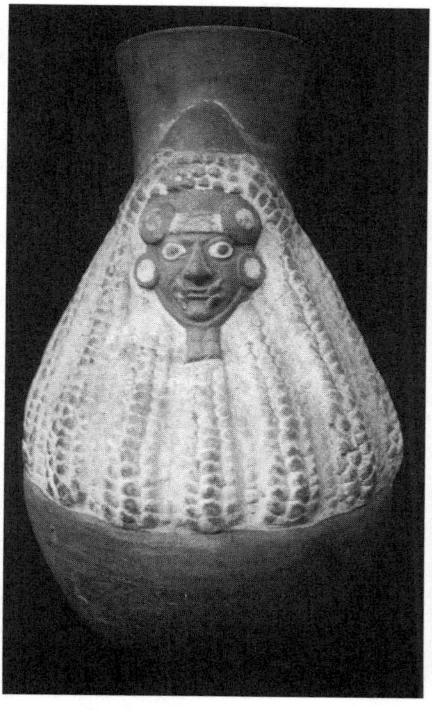

Fig. 29. P-19. Plain jar depicting an anthropomorphic figure with attributes of Ai-apaec.

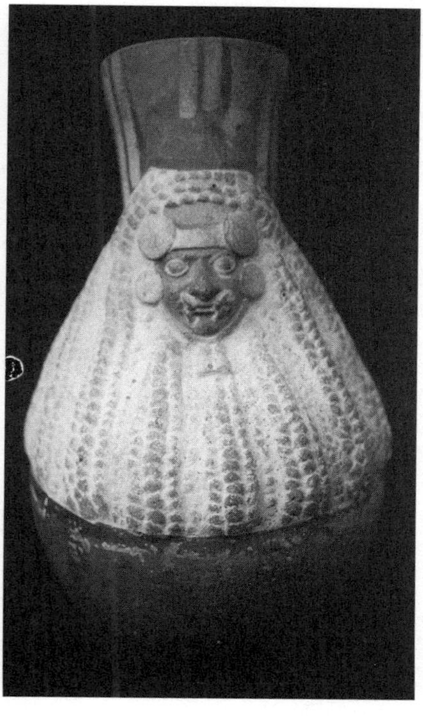

Fig. 30. P-20. Plain jar depicting an anthropomorphic figure with attributes of Ai-apaec.

Plain Jars Depicting an Anthropomorphic Figure with Attributes of Ai-apaec Accompanied by Two Companion Figures

P-21. Figs. 31 and 32. Plain jar depicting an anthropomorphic figure with attributes of Ai-apaec accompanied by two companion figures.
Provenience: excavated in the Lambayeque Valley.
Museo Nacional de Antropología y Arqueología KK/16624.
Fabric: red and white.
Dimensions: ht. 31.0 cm, width 18.4 cm.
Race of maize: Pira/Nal Tel.

The vessel has an unadorned globular body. The slightly flaring neck of the jar emerges from behind the head of the figure of Ai-apaec. The figure has the wide, partially open mouth with fangs and round eyes displayed on other jars with a provenience of the Lambayeque Valley. It also has strands

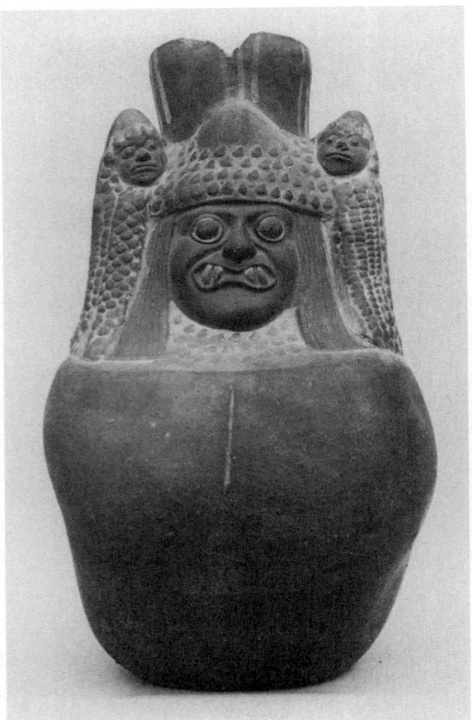

Fig. 31. P-21. Plain jar depicting an anthropomorphic figure with attributes of Ai-apaec accompanied by two companion figures.

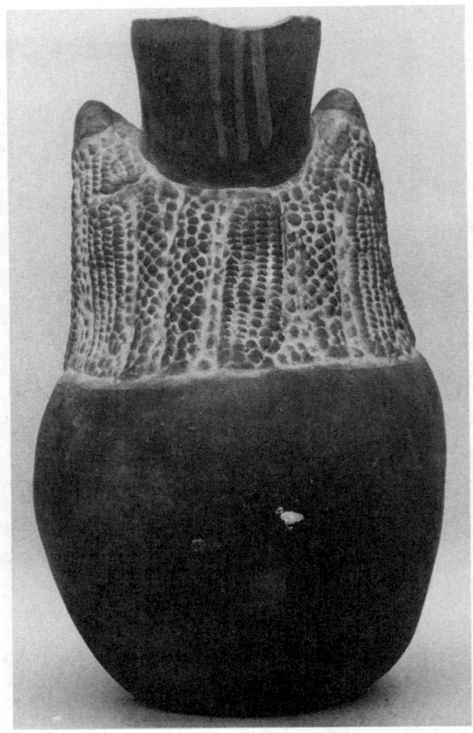

Fig. 32. P-21. Rear view of fig. 31.

of hair hanging down the sides of its face to about shoulder length. It appears to be wearing a pointed cap, bare at the tip and otherwise covered by pointed kernel-like elements, and the same kernel elements form a collar around the figure's neck. There is an ear of maize on each side of the head, and emerging from the tip of each maize ear is the face of a small human-like figure without fangs. The heads of the two small figures also end in a pointed terminus. These jars with small companion figures may symbolize maize introgressed with *Tripsacum*, a wild grass related to maize. Such plants produce extra little companion ears on the same stalk basal to the main ear, which emerges first. Surrounding the jar behind the head of Ai-apaec are more maize ears. The ears alternate between straight and spiraling kernel rows (fig. 32). The neck of the jar is decorated by four groups of three parallel vertical white lines.

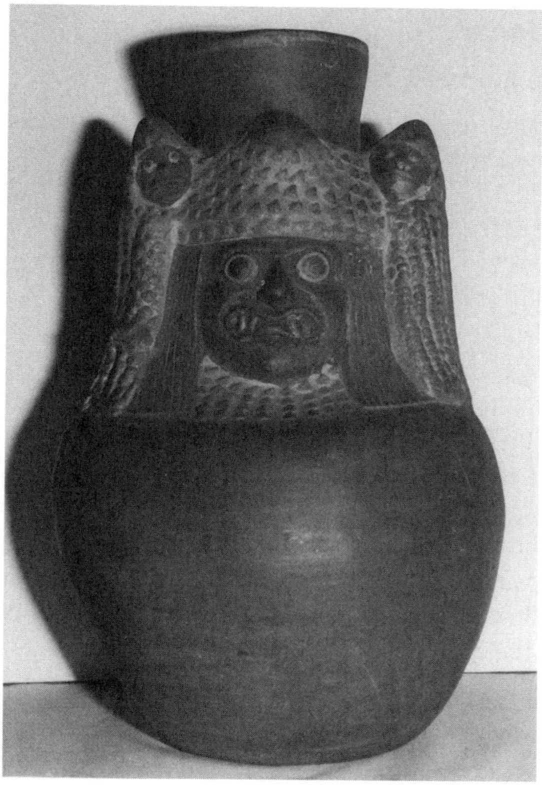

Fig. 33. P-22. Plain jar depicting an anthropomorphic figure with attributes of Ai-apaec accompanied by two companion figures.

P-22. Fig. 33. Plain jar depicting an anthropomorphic figure with attributes of Ai-apaec accompanied by two companion figures.
Provenience: Pacasmayo.
Brooklyn Museum of Art 1890.

Fabric: red and white.
Dimensions: ht. 31.0 cm, width 20.0 cm.
Race of maize: Pira/Nal Tel.

This jar is virtually identical to the one described under P-21. The only difference is that it does not have the vertical parallel stripes on the neck. Instead, a white stripe encircles the rim of the mouth of the jar. Another similar vessel is illustrated by Arthur Baessler (1902–3, fig. 292).

P-23. Plain jar depicting an anthropomorphic figure with attributes of Ai-apaec accompanied by two companion figures.
Provenience: Pacasmayo.
Museum für Völkerkunde Berlin 17445.
Fabric: red.
Dimensions: ht. 23.8 cm, width 18.8 cm.
Race of maize: Pira/Nal Tel.

This jar is similar to the ones described in P-21 and P-22.

P-24. Figs. 34 and 35. Plain jar depicting an anthropomorphic figure with attributes of Ai-apaec accompanied by two companion figures.
Provenience: unknown.
Museo Rafael Larco Herrera 2300.
Fabric: red and white.
Race of maize: Pollo.

On this jar, the anthropomorphic figure wears a bead necklace and snake-head ear ornaments and is accompanied by two smaller, fangless figures. All the figures have bare pointed heads. The jar is redware with a plain lower section surmounted by eight distinct ears of maize encircling the vessel. A companion figure emerges from one of the ears on each side of Ai-apaec, and an arm of each figure reaches out to touch the maize ear before it. The background area of the maize and companion figures, including the head of the main figure, is painted white. The maize ears on this jar are in low relief, and presumably they do not depict full half rounds of the ear. Six evenly spaced ears encircle the jar, three in front and three in back.

P-25. Fig. 36. Plain jar depicting an anthropomorphic figure with attributes of Ai-apaec accompanied by two companion figures.
Provenience: unknown.
Brooklyn Museum of Art 41.1275–74.
Fabric: red, black, and white.
Dimensions: ht. 22.0 cm, width 13.0 cm.
Race of maize: Confite Morocho.

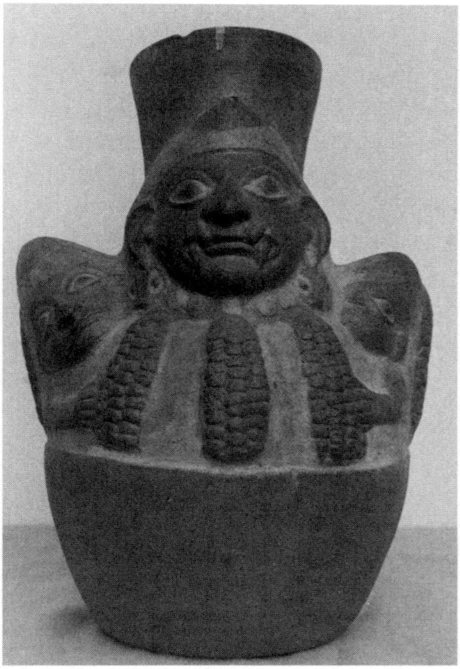

Fig. 34. P-24. Plain jar depicting an anthropomorphic figure with attributes of Ai-apaec accompanied by two companion figures.

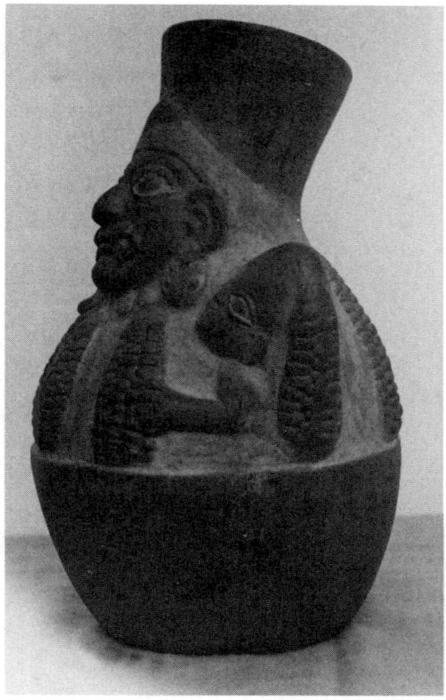

Fig. 35. P-24. Side view of fig. 34.

The jar for P-25 (fig. 36) represents an anthropomorphic figure associated with attributes of Ai-apaec accompanied by two companion figures. It is decorated with geometric designs painted in black and white. On the lower section of the jar, which has no relief decoration, there are circles and stepped-fret designs painted in black and white stripes. Above this area are six maize ears in relief around the jar with the intermediate areas decorated with large bosses. The fanged Ai-apaec figure is flanked by companion figures whose heads emerge from an ear of maize on each side. The heads of all three figures end in a bare, pointed terminus with a fillet of bosses across the forehead. The black geometric designs are characteristic of influence from the antecedent Gallinazo culture in the Virú Valley.

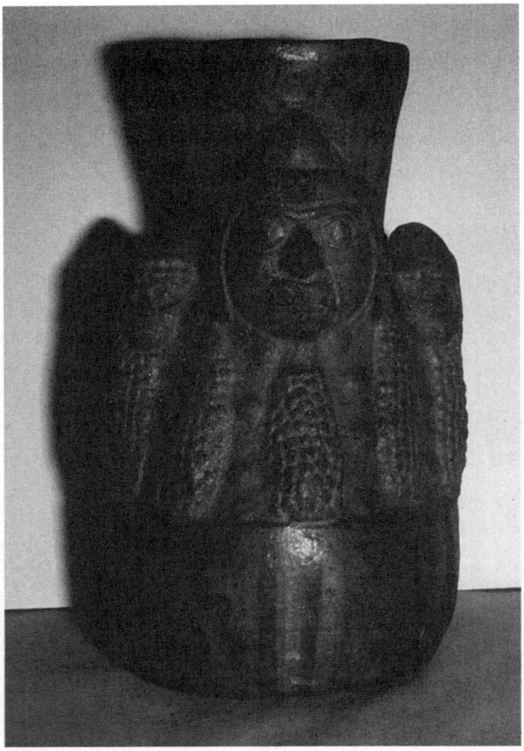

Fig. 36. P-25. Plain jar depicting an anthropomorphic figure with attributes of Ai-apaec accompanied by two companion figures.

Stirrup-Spout Jars Depicting an Anthropomorphic Figure with Attributes of Ai-apaec Accompanied by Two Companion Figures

P-26. Figs. 37 and 38. Stirrup-spout jar depicting an anthropomorphic figure with attributes of Ai-apaec accompanied by two companion figures.
Provenience: excavated in Huanchaco.
Museo Rafael Larco Herrera J-30839.
Fabric: red, black, and white.
Race of maize: Confite Morocho.

The spout on this jar depicting an anthropomorphic figure accompanied by two smaller figures is a modified form of the traditional stirrup spout and is diagnostic of Moche IVA (see fig. 7). The tubular spout is attached directly to the jar with a curved tubular handle attached to the body of the vessel and to the spout. This vessel also has an annular base. The Ai-apaec figure has rounded eyes, wide mouth with fangs, and whiskers painted in black on the cheeks. The companion figures flanking each side of the central figure have the same fangs and whisker markings but do not have ears showing, as does Ai-apaec. The lower part of the jar is plain, surmounted by the figures whose heads emerge from the body of the vessel. The jar is completely covered with an embossed background that extends up around the heads of the figures, which terminate in bare points. Three ears of maize are on the front of the vessel. The central ear is larger than the ears on each side of it. An arm of each companion figure is reaching around to the small ears. Around the mouth of the spout are black curvilinear designs that are reminiscent of designs on other Moche vessels, representing the tentacles of an octopus.

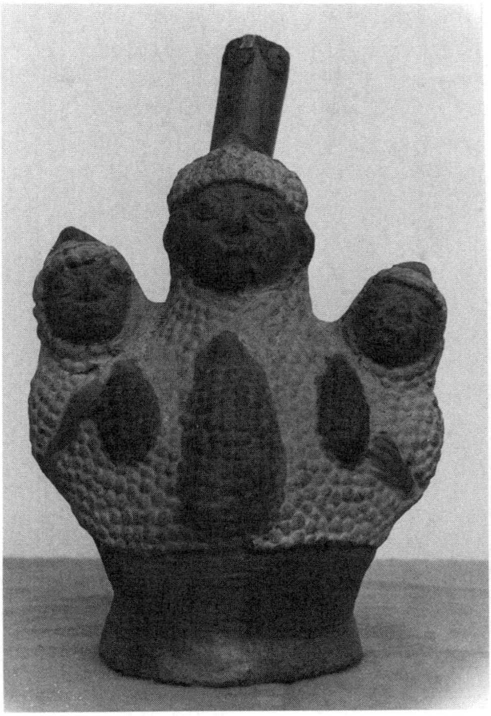

Fig. 37. P-26. Stirrup-spout jar depicting an anthropomorphic figure with attributes of Ai-apaec accompanied by two companion figures.

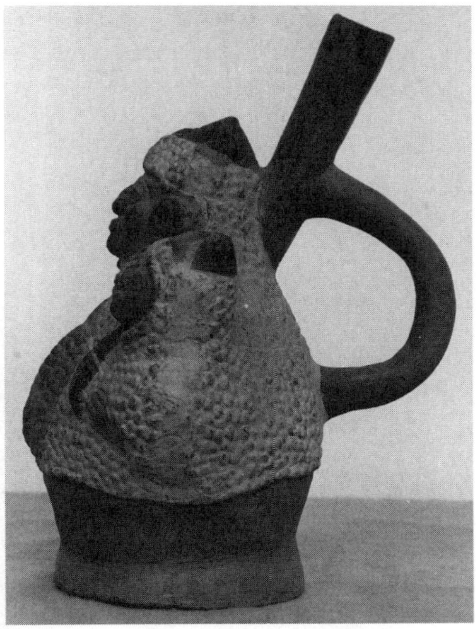

Fig. 38. P-26. Side view of fig. 37.

P-27. Fig. 39. Stirrup-spout jar depicting an anthropomorphic figure with attributes of Ai-apaec accompanied by two companion figures.
Provenience: Chimbote.
Museum für Völkerkunde Berlin 4652.
Fabric: red and white.
Dimensions: ht. 20.0 cm, width 15.6 cm.
Race of maize: Chaparreño.

The stirrup spout on this jar is broken off, but the two holes in back indicate that it was the traditional form of stirrup spout. The anthropomorphic figure has large flaring ears with snake-head ear ornaments and wears a pointed cap that is plain with a border of kernel-like elements. There are smaller figures without fangs with plain pointed caps and arms reaching out to hold onto the maize that completely surrounds the main portions of the jar. Whereas the headdresses of the figures are barren of kernels, the kernels on the mold-made maize ears encircling the jar cover the tips.

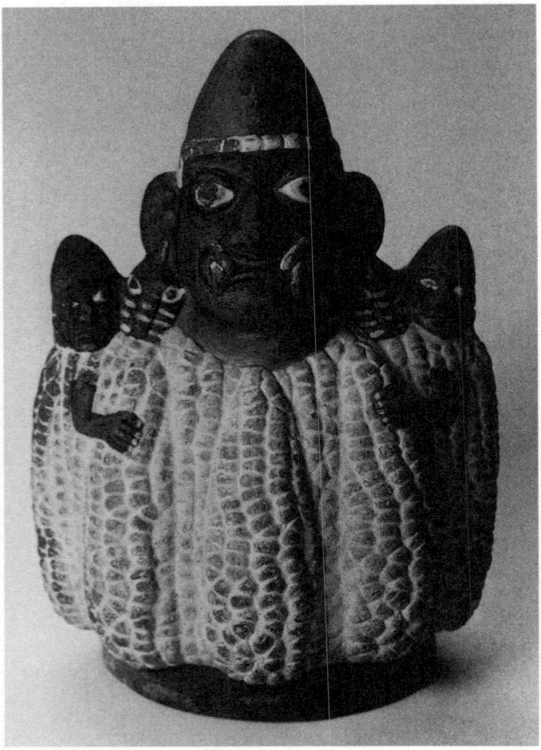

Fig. 39. P-27. Stirrup-spout jar depicting an anthropomorphic figure with attributes of Ai-apaec accompanied by two companion figures.

P-28. Stirrup-spout jar depicting an anthropomorphic figure with attributes of Ai-apaec accompanied by two companion figures.

Provenience: unknown.
American Museum of Natural History B8978.
Fabric: red and white.
Dimensions: ht. 22.5 cm, width 15.0 cm.
Race of maize: Pira/Nal Tel.

This jar is virtually identical in style to P-27 (fig. 39). The main figure has large flaring ears with snake-head ear ornaments and wears a plain pointed cap with a border of kernel-like elements. There are smaller figures without fangs with plain pointed caps and arms reaching out to hold onto the maize that completely surrounds the main portions of the jar. Whereas the head-dresses of the figures are barren of kernels, the kernels on the mold-made maize ears encircling the jar cover the tips.

P-29. Fig. 40. Stirrup-spout jar depicting an anthropomorphic figure with attributes of Ai-apaec accompanied by two companion figures.
Provenience: unknown.
Museo Rafael Larco Herrera 3993.
Fabric: black.
Race of maize: Pira/Nal Tel.

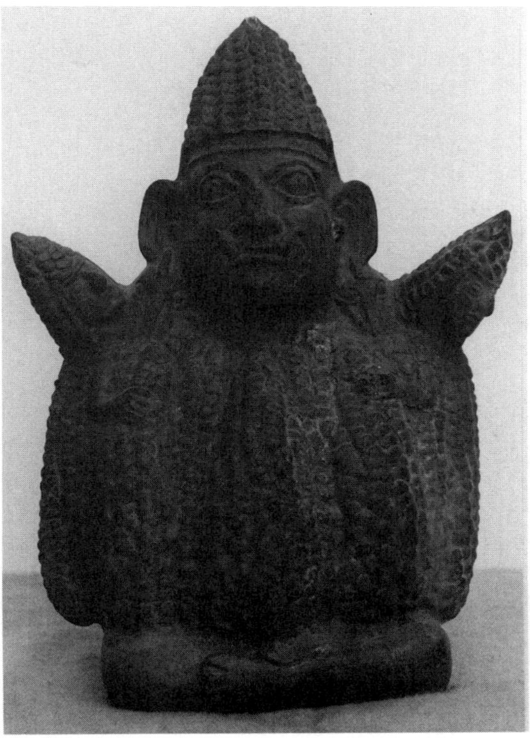

Fig. 40. P-29. Stirrup-spout jar depicting an anthropomorphic figure with attributes of Ai-apaec accompanied by two companion figures.

The stirrup spout on this jar is broken off, but the two holes in back indicate that it was the traditional form of stirrup spout. A tiny hole in the top of the cap of the central figure indicates that this was a whistling jar. The primary figure has large flaring ears with snake-head ear ornaments and wears a pointed cap covered with pointed kernel-like elements. There are smaller figures without fangs that have pointed caps covered with kernels and arms reaching out to hold onto the maize that completely surrounds the main portions of the jar. The base of the jar shows Ai-apaec seated cross-legged. This is the only instance among the Moche vessels of one of these figures seated cross-legged. It is, however, the prevalent posture for figures on Zapotec urns. Whereas the headdresses of the figures are completely covered with kernels, the mold-made maize ears encircling the jar have barren tips.

P-30. Fig. 41. Stirrup-spout jar depicting an anthropomorphic figure with attributes of Ai-apaec accompanied by two companion figures.
Provenience: unknown.
Yale University Peabody Museum of Natural History 235134.
Fabric: red and white.

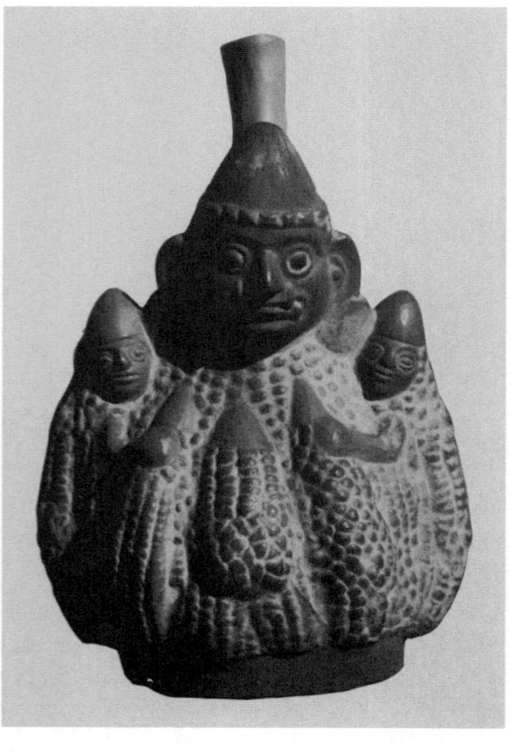

Fig. 41. P-30. Stirrup-spout jar depicting an anthropomorphic figure with attributes of Ai-apaec accompanied by two companion figures.

Dimensions: ht. 24.0 cm, width 15.0 cm.
Race of maize: Pira/Nal Tel.

The primary figure's head ends in a barren pointed terminus with a border of a single row of kernel-like elements at the base around the forehead. It has fangs, but the two smaller companion figures do not. Each one has an arm reaching around the front holding onto an ear of maize. The main section of the jar is modeled from naturalistic mold-made maize ears. The bottom of the jar is plain.

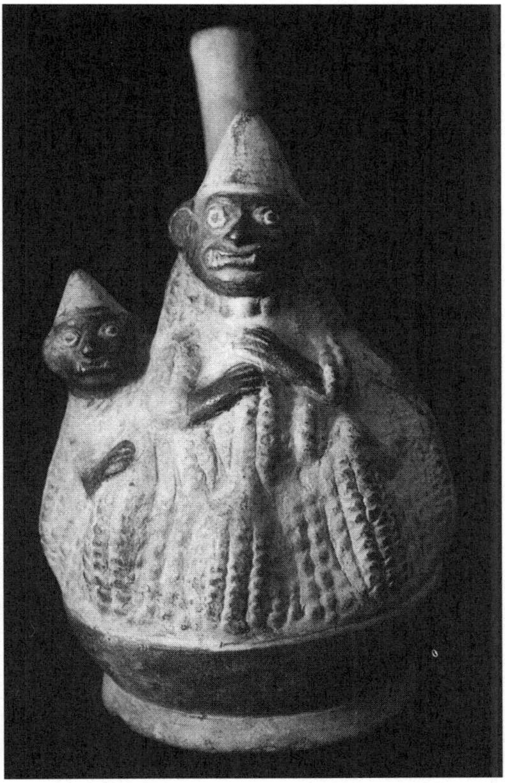

Fig. 42. P-31. Stirrup-spout jar depicting an anthropomorphic figure with attributes of Ai-apaec accompanied by two companion figures.

P-31. Fig. 42. Stirrup-spout jar depicting an anthropomorphic figure with attributes of Ai-apaec accompanied by two companion figures.
Provenience: unknown.
American Museum of Natural History 41.2/7431.
Fabric: red and white.
Dimensions: ht. 22.5 cm, width 15.0 cm.
Race of maize: Pira/Nal Tel.

The anthropomorphic figure's head and the head of the companion figure end in a barren, pointed terminus. This jar is unusual because it depicts a single companion figure rather than two, and the small figure has fangs like the main figure's. Both figures hold onto the maize that covers the body of the jar. The lower section of the jar is plain, and the base is annular. Ai-apaec is adorned with a bead necklace.

Stirrup-Spout Jars Depicting an Anthropomorphic Figure with Attributes of Ai-apaec

P-32. Figs. 43 and 44. Stirrup-spout jar depicting an anthropomorphic figure with attributes of Ai-apaec.
Provenience: excavated in the Virú Valley.
Museo Nacional de Antropología y Arqueología 1/2865.
Fabric: red and white.
Dimensions: ht. 19.4 cm, width 18.4 cm.
Race of maize: Chococeño.

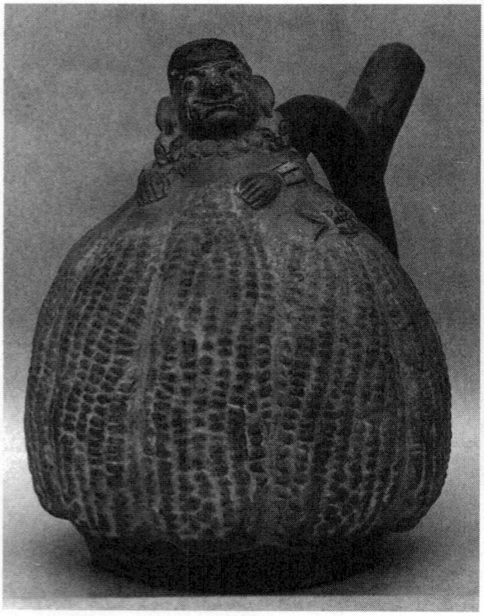

Fig. 43. P-32. Stirrup-spout jar depicting an anthropomorphic figure with attributes of Ai-apaec.

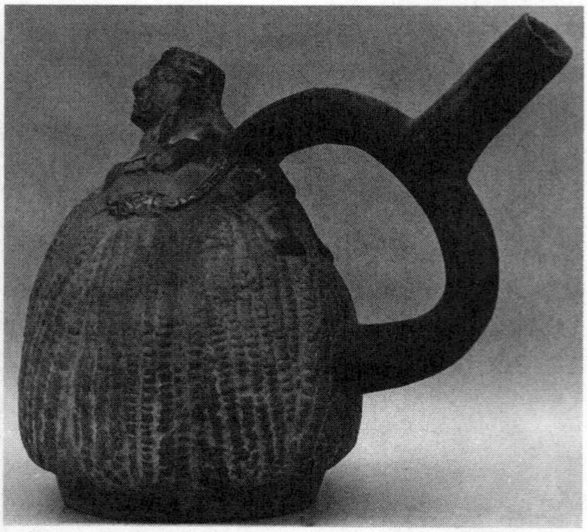

Fig. 44. P-32. Side view of fig. 43.

The body of the vessel is entirely surrounded by large, thick conical ears of maize, surmounted by an anthropomorphic figure lying on his stomach with arms in front and legs spread out in back. Part of the spout is attached to the figure's back. The figure is accompanied by a snake curling around the jar. The figure has large round eyes, a very wide fanged mouth, and large ears with snake-head ear ornaments. He wears a bead necklace and red and white striped bands around his wrists. His clothing is painted red, and he has bare feet. He wears a fillet headdress that originally had an ornament attached to the front. The broken ornament may have been a feline head, which is often seen on other depictions of Ai-apaec wearing a fillet headdress (Kroeber 1925, 220).

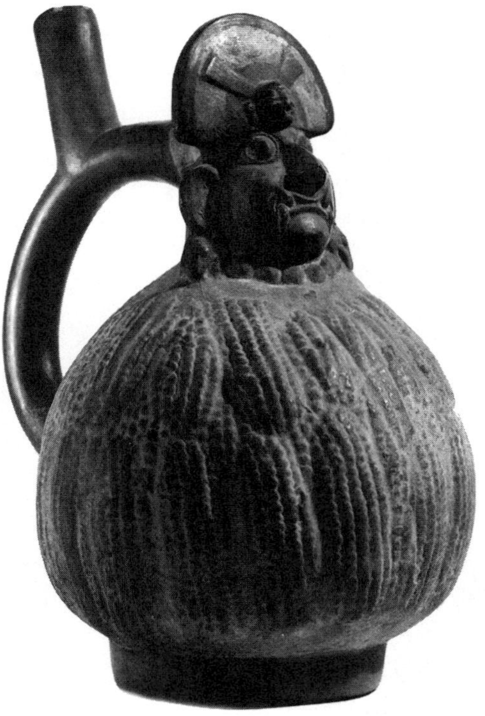

Fig. 45. P-33. Stirrup-spout jar depicting an anthropomorphic figure with attributes of Ai-apaec.

P-33. Fig. 45. Stirrup-spout jar depicting an anthropomorphic figure with attributes of Ai-apaec.
Provenience: unknown.
Musée de l'Homme 87.134.14.
Fabric: red.
Dimensions: ht. 23.1 m, width 21.3 m.
Race of maize: Pira/Nal Tel.

The body of the vessel is covered with ears of maize in low relief, surmounted by a figure of Ai-apaec as a warrior. Ai-apaec has large round eyes, a fanged mouth, large ears with snake-head ear ornaments, and a bead necklace. He wears a sunburst headdress with a feline head in front. There are black painted lines on the nose and eyes.

P-34. Fig. 46. Stirrup-spout jar depicting an anthropomorphic figure with attributes of Ai-apaec.
Provenience: Chimbote.
Chicago Field Museum of Natural History 1229.
Fabric: red and white.
Dimensions: ht. 22.5 cm, width 14.0 cm.
Race of maize: Pira/Nal Tel.

The stirrup-spout jar is covered with ears of maize surmounted by a figure of Ai-apaec as a warrior. Ai-apaec has large round eyes, a fanged mouth,

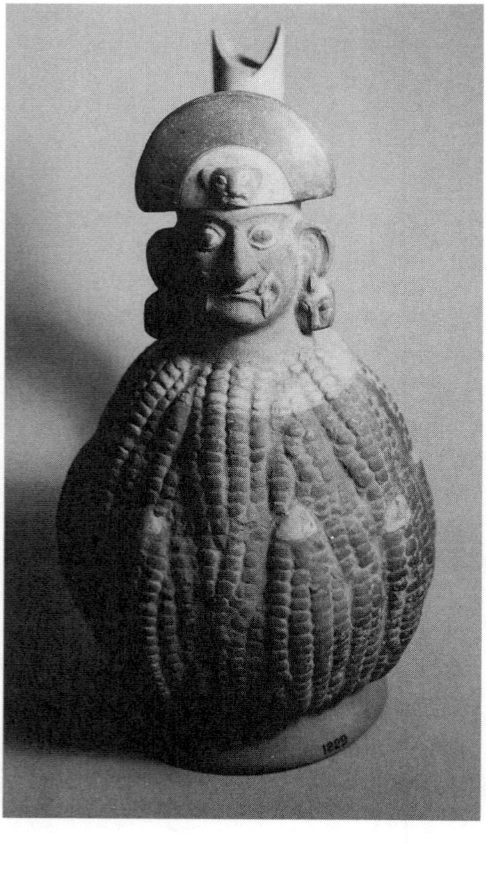

Fig. 46. P-34. Stirrup-spout jar depicting an anthropomorphic figure with attributes of Ai-apaec.

and snake-head ear ornaments. He wears a sunburst headdress with a feline head in front. The stirrup spout, barren tips of the maize ears, fangs, and sclera of the eyes are accented with white paint.

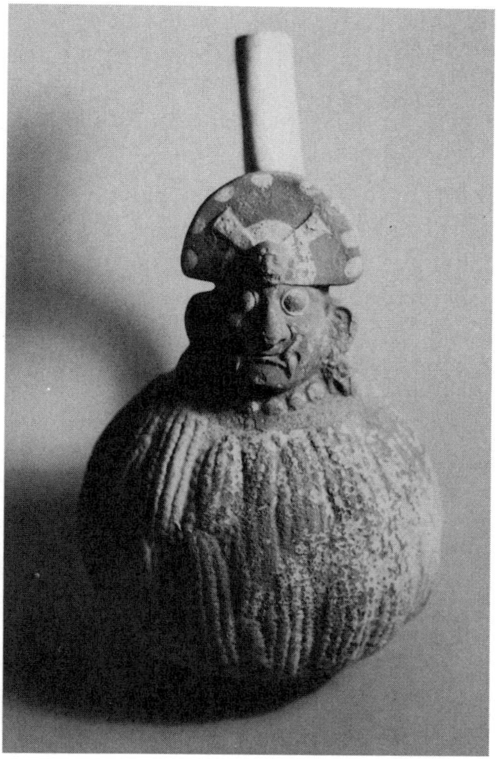

Fig. 47. P-35. Stirrup-spout jar depicting an anthropomorphic figure with attributes of Ai-apaec.

P-35. Fig. 47. Stirrup-spout jar depicting an anthropomorphic figure with attributes of Ai-apaec.
Provenience: Chimbote.
Chicago Field Museum of Natural History 4794.
Fabric: red and white.
Dimensions: ht. 23.3 cm, width 22.0 cm.
Race of maize: Pira/Nal Tel.

The body of the vessel is covered with ears of maize in low relief, surmounted by a figure of Ai-apaec as a warrior. The figure has large, round eyes, a fanged mouth, snake-head ear ornaments, and a bead necklace. He wears a sunburst headdress with a feline head in front and white circular designs. The stirrup spout, maize ears covering the body of the jar, fangs, and sclera of the eyes are accented with white paint.

P-36. Fig. 48. Stirrup-spout jar depicting an anthropomorphic figure with attributes of Ai-apaec.
Provenience: unknown.
University of Pennsylvania, University Museum 67–36–1.
Fabric: red and white.
Dimensions: ht. 22.4 cm, width 15.2 cm.
Race of maize: Chapalote/Nal Tel.

The technical execution of this jar is excellent. It is beautifully formed, highly polished, and well fired. The anthropomorphic figure wears the characteristic helmet of a Moche warrior, and the kernel elements cover the head above the helmet features. The head forms a bare pointed terminus. Ai-apaec's arms are shown reaching out to hold onto two maize ears in front of the jar, and the vessel is surrounded by ten maize ears.

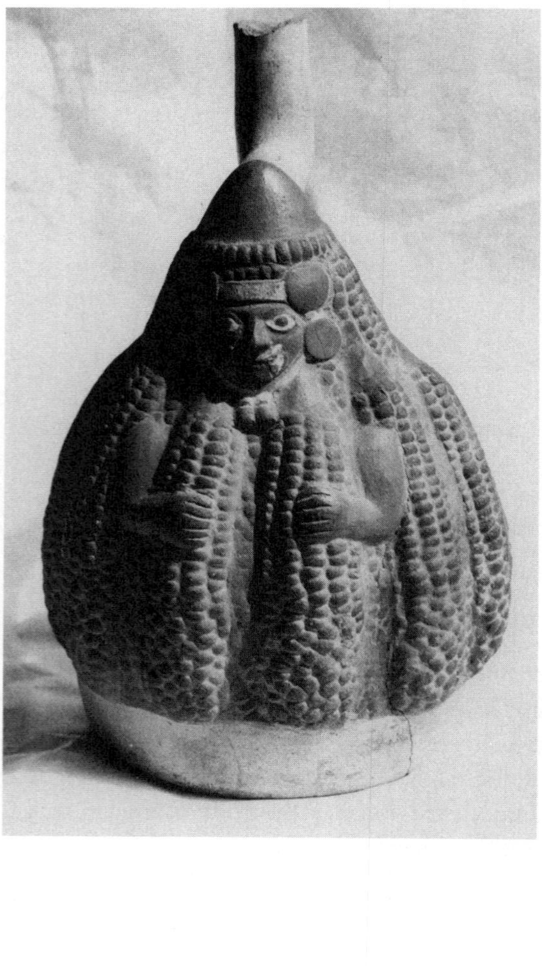

Fig. 48. P-36. Stirrup-spout jar depicting an anthropomorphic figure with attributes of Ai-apaec.

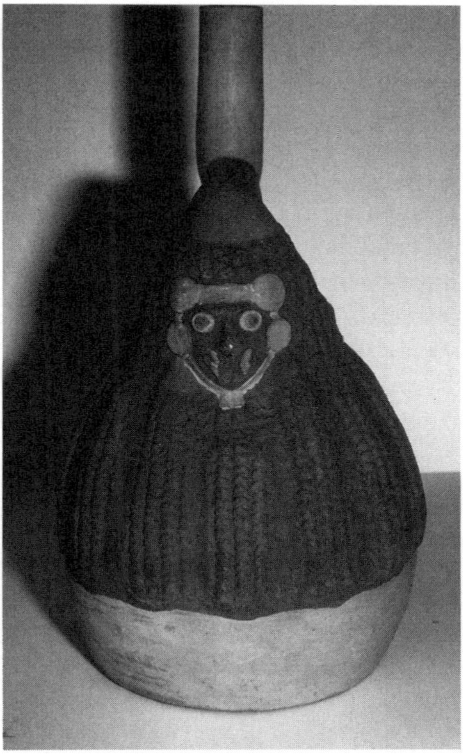

Fig. 49. P-37. Stirrup-spout jar depicting an anthropomorphic figure with attributes of Ai-apaec.

P-37. Fig. 49. Stirrup-spout jar depicting an anthropomorphic figure with attributes of Ai-apaec.
Provenience: unknown.
Museum of the American Indian, Heye Foundation 5/1741.
Fabric: red and white.
Dimensions: ht. 19.0 cm, width 13.8 cm.
Race of maize: Confite Morocho.

This jar is similar to the one described for P-36. The plain lower portion of the redware vessel and the stirrup spout are painted white, as are the features of the Moche warrior's helmet, the bare tip of the head, the sclera of the eyes, and the fangs.

P-38. Stirrup-spout jar depicting an anthropomorphic figure with attributes of Ai-apaec.
Provenience: Chimbote.
Museum für Völkerkunde Berlin 17827.
Fabric: red and white.

Dimensions: ht. 20.0 cm, width 13.8 cm.
Race of maize: Confite Morocho.

The figure on this jar has a completely plain, pointed headdress that resembles a warrior's helmet without the typical round motifs at each side above the forehead as on figures P-36 and P-37 but wears the round earspools and has the helmet neck strap characteristic of a warrior. It is different, however, because it does not have fangs. The top of the pointed helmet is painted white. The transition between the plain lower portion of the redware vessel and the upper body is delineated by a white band. The stirrup spout has traces of white paint. The sclera of the eyes and the fangs, the neck strap and earspools, and the spaces between the ears of maize are painted white. Eleven ears of maize surround the upper two-thirds of the vessel.

P-39. Fig. 50. Stirrup-spout jar depicting an anthropomorphic figure with attributes of Ai-apaec.
Provenience: unknown.
Museo Rafael Larco Herrera 2281.
Fabric: red.
Race of maize: Pira/Nal Tel.

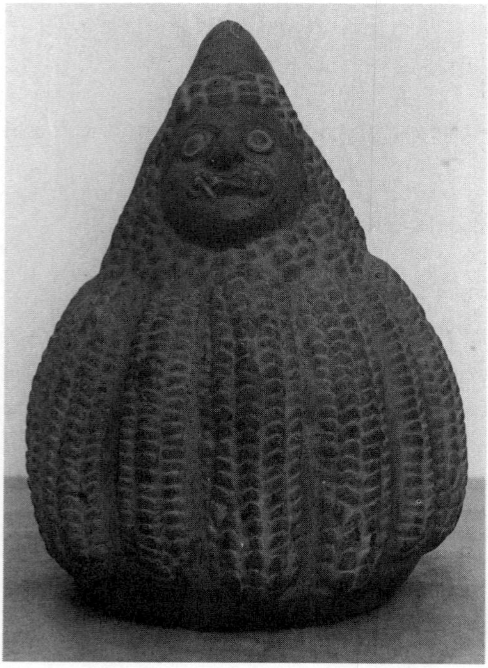

Fig. 50. P-39. Stirrup-spout jar depicting an anthropomorphic figure with attributes of Ai-apaec.

There is a plain narrow band at the base of this vessel. The remainder of the jar is covered by ears of maize. Only the fanged face of the figure and the bare pointed terminus of the head are displayed. Thirteen ears of maize surround the jar. The maize is identified as Pira/Nal Tel based on the statistical match of the morphometric measurements, but the slightly pointed kernels are more like Confite Morocho.

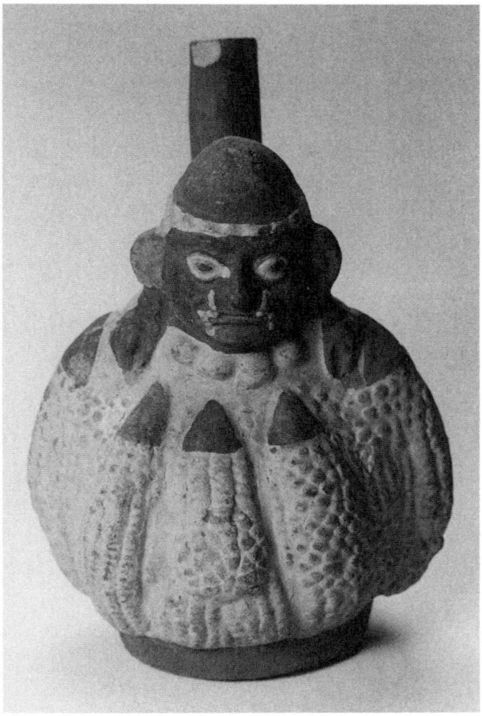

Fig. 51. P-40. Stirrup-spout jar depicting an anthropomorphic figure with attributes of Ai-apaec.

P-40. Fig. 51. Stirrup-spout jar depicting an anthropomorphic figure with attributes of Ai-apaec.
Provenience: Chimbote.
Museum für Völkerkunde Berlin 17826.
Fabric: red and white.
Dimensions: ht. 19.4 cm, width 14.4 cm.
Race of maize: Pira/Nal Tel.

This fanged anthropomorphic figure has a pointed head with bare terminus delineated by kernel elements surrounding the edge along the forehead. It is adorned with a bead necklace and snake-head ear ornaments. The vessel is redware with white paint on the fangs, sclera of the eyes, bead necklace, and all the kernels of the ears of maize, and other kernel elements. White lines are painted around the rim of the spout and the handle. The figure's head is tilted forward like that of figure 16 (P-8).

P-41. Fig. 52. Stirrup-spout jar depicting an anthropomorphic figure with attributes of Ai-apaec.
Provenience: Chimbote.
Chicago Field Museum of Natural History 4796.
Fabric: red and white.
Dimensions: ht. 22.0 cm, width 21.0 cm.
Race of maize: Pira/Nal Tel.

This fanged Ai-apaec figure has a pointed head with bare terminus. It is adorned with a bead necklace and snake-head ear ornaments. The vessel is redware with white paint on the fangs, sclera of the eyes, bead necklace, ears of maize, and annular ring at the base.

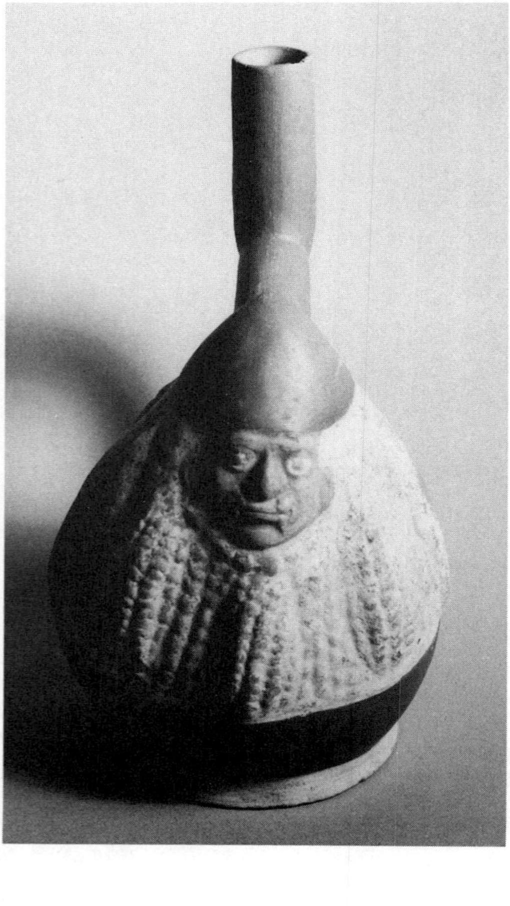

Fig. 52. P-41. Stirrup-spout jar depicting an anthropomorphic figure with attributes of Ai-apaec.

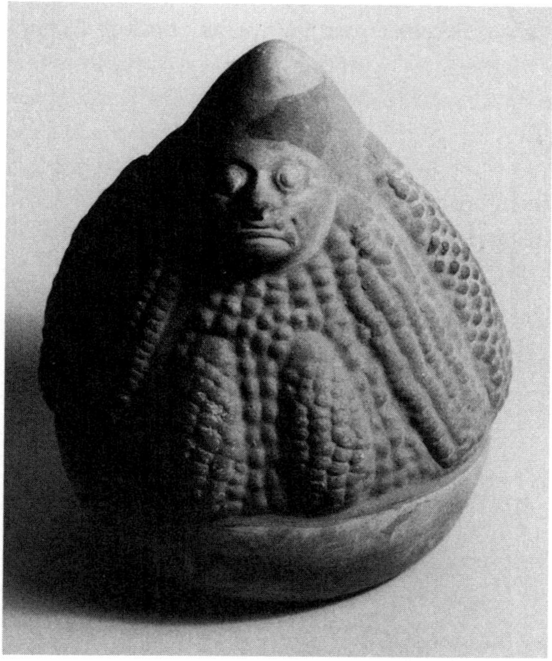

Fig. 53. P-42. Stirrup-spout jar depicting an anthropomorphic figure with attributes of Ai-apaec.

P-42. Fig. 53. Stirrup-spout jar depicting an anthropomorphic figure with attributes of Ai-apaec.
Provenience: Santa Valley.
Chicago Field Museum of Natural History 5667.
Fabric: redware.
Dimensions: ht. 18.0 cm, width 16 cm.
Race of maize: Pira/Nal Tel and Confite Puneño/Quicheño Rojo.

This fanged figure has a pointed head with bare terminus. The redware jar has white paint on the fangs and sclera of the eyes and incised old-age lines on the face. There are three types of maize: small, grenade-shaped ears flanked by long thin ears with straight rows and long thin ears with irregular rows.

P-43. Stirrup-spout jar depicting an anthropomorphic figure with attributes of Ai-apaec.
Provenience: unknown.
Museum of the American Indian, Heye Foundation 4/8898.
Fabric: red and white.
Dimensions: ht. 20.6 cm, width 11.9 cm.
Race of maize: Pira/Nal Tel.

The spout of this jar deviates somewhat from the forms of other stirrup-spout jars. The tubular spout emerging from the vessel is unusually wide. A flattened, curved handle is attached to the spout and to the body of the vessel. Unlike other Moche IVA jars of this type, this handle is not a hollow tube. The vessel is redware with white paint on the spout, the narrow base, the bare pointed head, the round earspools the figure is wearing, the sclera of the eyes, and the fangs. The figure's head is tilted forward like that of figure 16 (P-8).

P-44. Fig. 54. Stirrup-spout jar depicting an anthropomorphic figure with attributes of Ai-apaec.
Provenience: unknown.
American Museum of Natural History B/5007.
Fabric: red and white.
Dimensions: ht. 21.3 cm, width 13.8 cm.
Race of maize: Pira/Nal Tel.

This jar closely resembles the one in P-43. The figure's head is tilted forward like those of figures 16 (P-8) and P-43. The vessel is redware with white

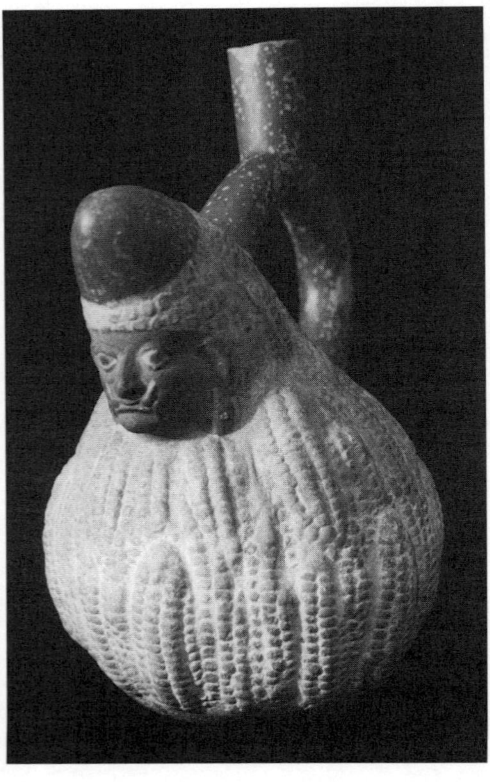

Fig. 54. P-44. Stirrup-spout jar depicting an anthropomorphic figure with attributes of Ai-apaec.

paint on the spout, the maize covering the body of the jar, the sclera of the eyes, and the fangs. The figure is wearing snake-head ear ornaments.

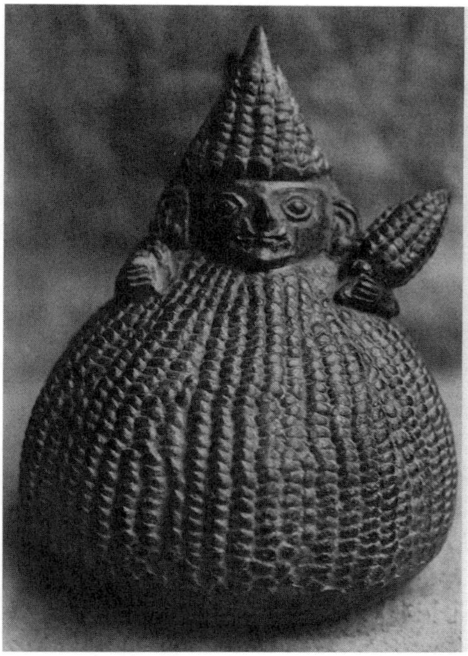

Fig. 55. P-45. Stirrup-spout jar depicting an anthropomorphic figure with attributes of Ai-apaec.

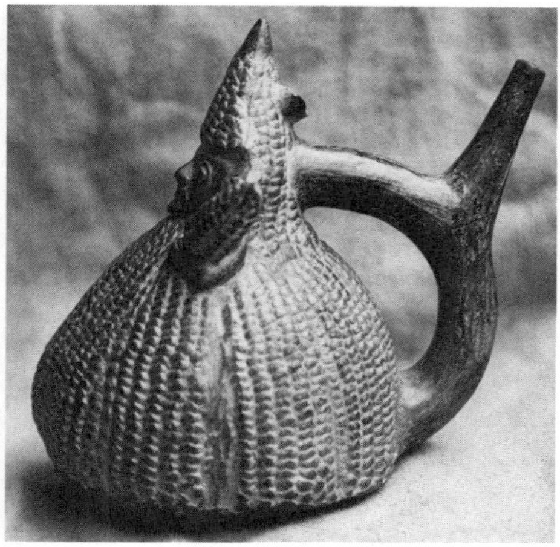

Fig. 56. P-45. Side view of fig. 55.

P-45. Figs. 55 and 56. Stirrup-spout jar depicting an anthropomorphic figure with attributes of Ai-apaec.
Provenience: unknown.
Museo Arqueológico Brüning de Lambayeque B-99.
Fabric: black.
Period: Chimú.
Race of maize: Confite Morocho.

This vessel is blackware diagnostic of the Chimú culture, which followed Moche cultural development on the North Coast after an interim period of stylistic influence and domination from Tiahuanaco. The Chimú culture dates to ca. A.D. 1100. This period saw the revival of such Moche traits as the stirrup-spout jar, the fanged god, and representations of maize in relief, some of which were still mold-made from real maize. All of these traits were suppressed during Tiahuanacoid domination. They were later revived in distinct, modified forms. Blackware was by this time the prevalent form of pottery, and the stirrup-spout jars have characteristic narrow spouts, as seen on figure 56. The wide, globular body of this vessel is completely covered with kernel elements which apparently were molded from actual maize. But the ears are so well blended into the body of the jar that they are indistinguishable and cannot serve for racial identification. Aiapaec wears a pointed cap covered with kernels, except for the extreme tip, which is bare. The figure has large ears and fangs, and its two hands are shown. A very small ear of maize is held in one hand. Something was held in the other hand, but it has been broken off.

Atypical Jars with Human Figures and Maize

P-46. Fig. 57. Stirrup-spout double jar of a human figure riding a maize boat.
Provenience: unknown.
Museum für Völkerkunde Berlin 17831.
Fabric: black.
Dimensions: ht. 12.5 cm, width 18.1 cm.
Race of maize: unidentified.

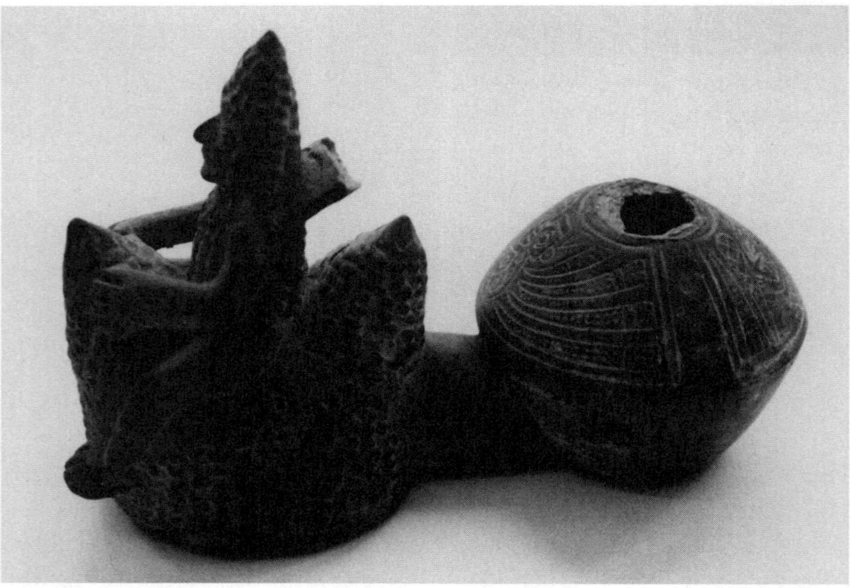

Fig. 57. P-46. Stirrup-spout double jar of a human figure riding a maize boat.

This double vessel joined by a stirrup-spout handle is the blackware diagnostic of the Chimú culture that followed the Moche style on the North Coast. One section of the jar depicts a human figure astride a boat completely covered with maize ears. The ears of maize are naturalistic, but the way they have been blended into the boat form obscures measurements necessary to identify the maize. This unusual piece may be transitional from the period of Tiahuanaco domination to Chimú because it combines the maize figure with the pointed head typical of Moche pottery with the later blackware, and the second container section of the jar has incised figures in the Tiahuanaco style. The spout that was attached to this section has been broken off, but a small part of the stirrup handle is still attached to the back of the human figure, indicating that it is a stirrup-spout jar.

P-47. Fig. 58. Stirrup-spout jar depicting a human figure.
Provenience: unknown.
Musées Royaux d'Art et d'Histoire AAM.39.107.
Fabric: black.
Dimensions: ht. 21.3 cm, width 16.9 cm.
Race of maize: Karapampa.

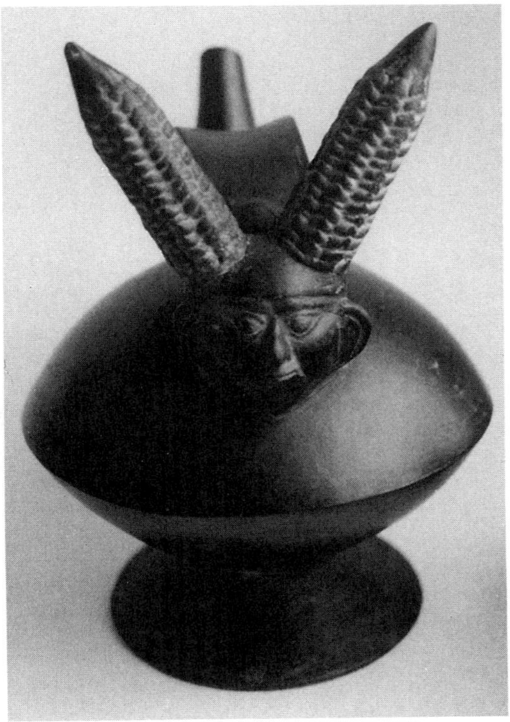

Fig. 58. P-47. Stirrup-spout jar depicting a human figure.

This black Chimú stirrup-spout jar with a globular body has a human face on the front with two ears of maize emerging from the top of its head like horns. The well-executed ears of maize are full round and have imbricated kernels and barren tips.

P-48. Figs. 59 and 60. Olla depicting a human figure.
Provenience: Chiclayo.
Musée de l'Homme 83.30.161.
Fabric: highly polished polychrome.
Dimensions: ht. 11.7 cm, width 14.4 cm.
Race of maize: Chapalote/Nal Tel.

This red clay olla painted with dark red on white is quite different from any other jar depicting maize I have seen. The jar is rounded and globular in shape. A human face with ears is modeled on the spout, which has an everted lip. The figure's arms and hands are extended around the front of the jar. The upper half of the front of the jar is painted white with a red checkered pattern representing the figure's garment. A zigzag pattern adorns the neck of the garment. On the back of the jar are two full-round ears of maize, and a red zigzag pattern encircles the midsection. The tip of one of the maize ears is broken, but the other one is intact and has a barren tip. The spout in back is painted white with red stripes.

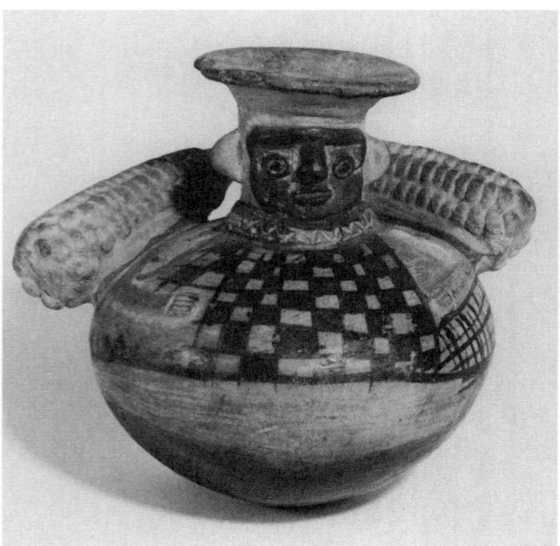

Fig. 59. P-48. Olla depicting a human figure.

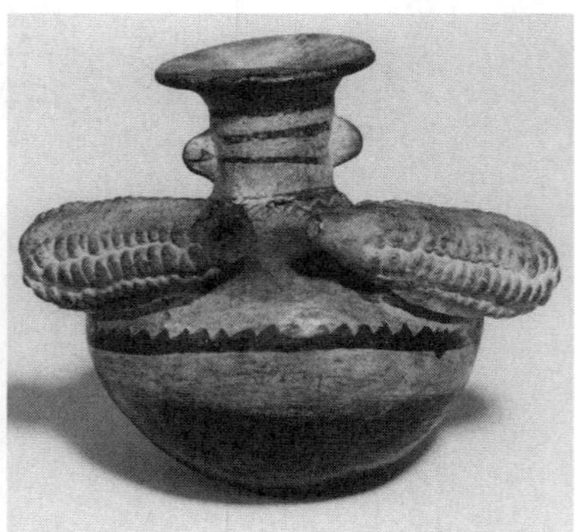

Fig. 60. P-48. Rear view of fig. 59.

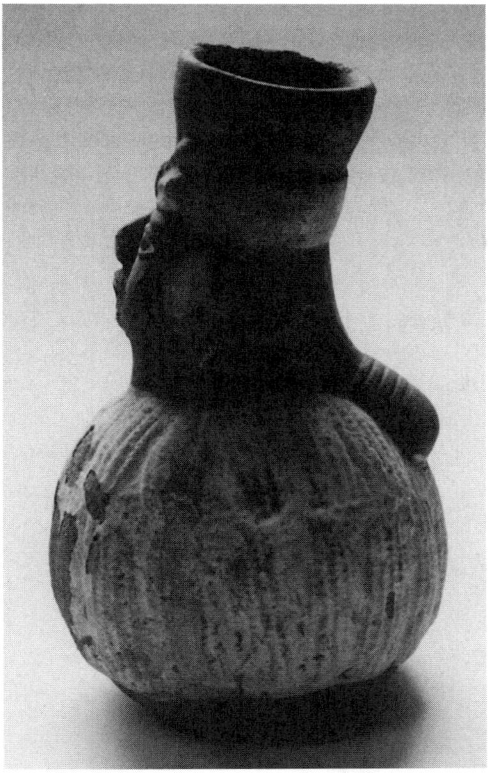

Fig. 61. P-49. Jar depicting a human figure.

P-49. Fig. 61. Jar depicting a human figure.
Provenience: Chimbote.
Museum für Völkerkunde Berlin 17820.
Fabric: red and white.
Dimensions: ht. 21.3 cm, width 13.8 cm.
Race of maize: Pira/Nal Tel.

The body of this globular jar is covered with ears of maize in low relief painted white. The long spout of the jar is shaped like a human head with large earspools projecting from it. The earspools have white crosses painted on them. The hair is pulled back in a ponytail, and the figure wears a turban tied in a bow in front. The sclera of the eyes and the turban are painted white.

P-50. Jar depicting a human figure.
Provenience: Chimbote.
Chicago Field Museum of Natural History 4792.
Fabric: red and white.

Dimensions: ht. 22.0 cm, width 21.0 cm.
Race of maize: unidentified.

This jar is similar in shape and type of maize in low relief as described for P-49. The figure wears round earspools and has a turban headdress tied in front with a bowknot, as does P-49. Since the jar was broken into pieces too fragmentary to measure when I saw it, no maize race identification could be determined, but since this jar is a virtual match to P-49, the race depicted is in all likelihood Pira.

P-51. Fig. 62. Olla depicting a human figure.
Provenience: unknown.
Museum für Völkerkunde Berlin 3304.
Fabric: black.
Dimensions: ht. 21.3 cm, width 13.8 cm.
Race of maize: Niñuelo.

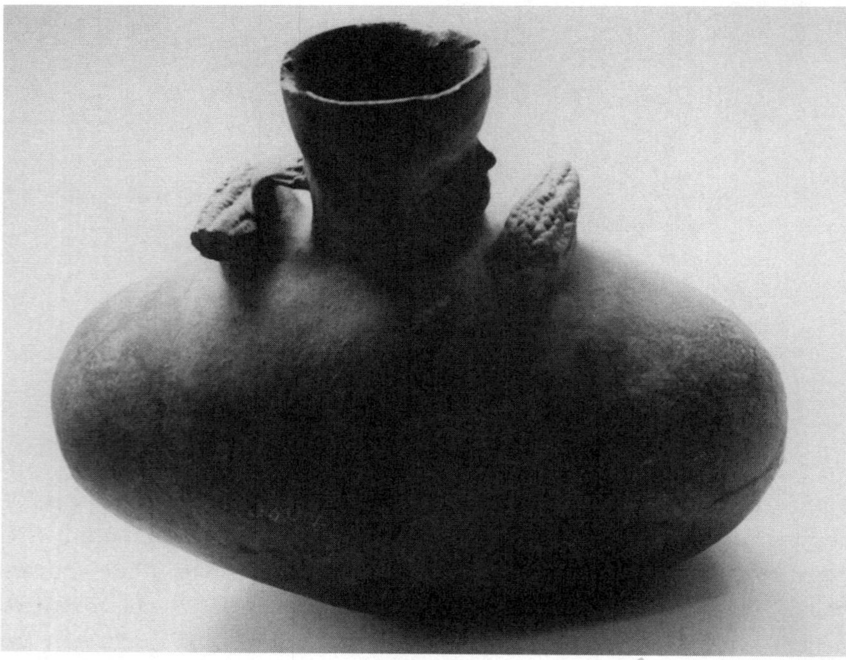

Fig. 62. P-51. Olla depicting a human figure.

This elongated, odd-shaped olla looks somewhat like a torpedo. It is plain blackware typical of the Chimú era. The spout is shaped like a human face, and it is incised with lines on the cheeks and corners of the eyes reminiscent of old-age wrinkles. Behind the head is a small handle. There are two ears of maize, one in front of the face and one in back adjoining the handle.

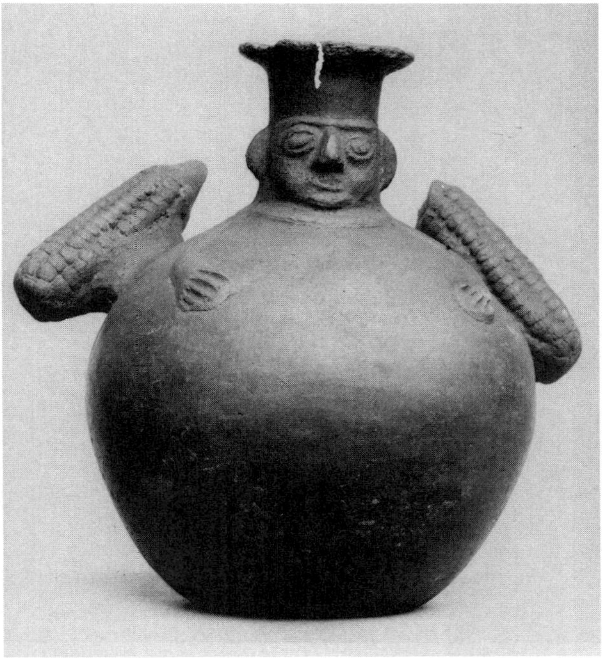

Fig. 63. P-52. Olla depicting a human figure.

P-52. Fig. 63. Olla depicting a human figure.
Provenience: unknown.
British Museum 1921. 10–27172.
Fabric: black.
Dimensions: ht. 17.5 cm, width 13.8 cm.
Race of maize: Pira/Nal Tel.

This Chimú olla represents a hunchbacked figure. The face is modeled on the spout of the jar. The figure wears a plain necklace, and its modeled arms extend around the front of the jar in low relief. On each upper side of the globular body of the jar is an attached ear of maize in full round that projects out from the jar.

Stirrup-Spout Jars Depicting Rodents

P-53. Fig. 64. Stirrup-spout jar depicting a rodent.
Provenience: excavated in the Virú Valley.
Museo Nacional de Antropología y Arqueología KK/16602.
Fabric: red and white.
Dimensions: ht. 19.5 cm, width 16.5 cm.
Race of maize: Choclero.

This is a Moche IVA redware stirrup-spout jar representing a rodent sitting up on its hind legs, holding a large ear of maize upright, and nibbling on the tip of the ear. White paint delineates the tail, whiskers, teeth, and ears of the animal and covers the vessel spout.

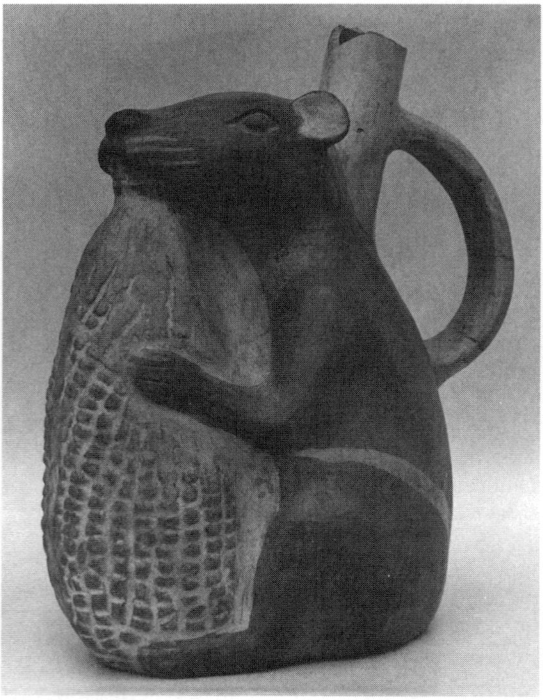

Fig. 64. P-53. Stirrup-spout jar depicting a rodent.

P-54. Fig. 65. Stirrup-spout jar depicting a rodent.
Provenience: unknown.
Duke University Museum of Art 73.1.507.
Fabric: red and white.
Dimensions: ht. 18.8 cm, width 17.5 cm.
Race of maize: Choclero.

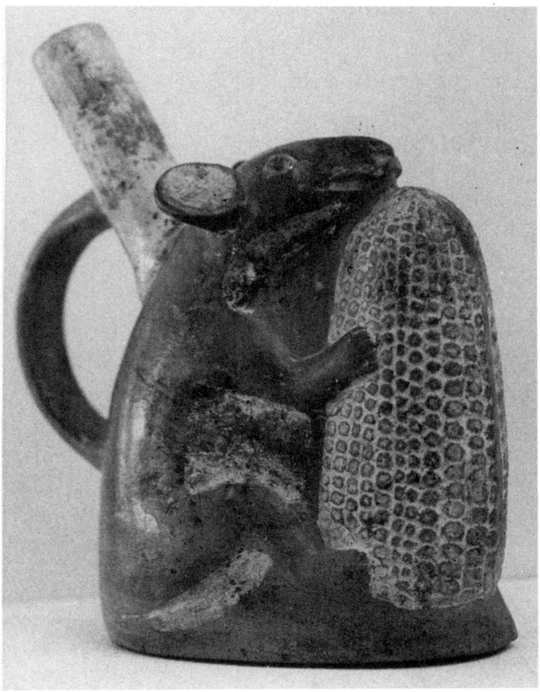

Fig. 65. P-54. Stirrup-spout jar depicting a rodent.

This is a Moche IVA redware stirrup-spout jar depicting a rodent similar to the one described in P-53. White paint decorates the animal's ears and abdomen, as well as the jar spout. The tail is modeled in low relief and is not painted white.

P-55. Stirrup-spout jar depicting a rodent.
Provenience: unknown.
Museum für Völkerkunde Berlin 2797.
Fabric: red and white.
Dimensions: ht. 21.3 cm, width 13.8 cm.
Race of maize: Choclero.

The stirrup spout on this jar is broken off, but the pieces are retained with this redware pot. The rodent and ear of maize resemble those in P-53 and P-54. The rodent is red. The maize ear, the inside of the figure's ears, sclera of the eyes, and spout are painted white. The maize is also highly similar in row number and large kernel size.

Olla Depicting a Frog with Maize and Manioc

P-56. Fig. 66. Olla depicting a frog with maize and manioc.
Provenience: Chimbote.
Museum für Völkerkunde Berlin 18175.
Fabric: red and white.
Dimensions: ht. 18.8 cm, width 25.6 cm.
Race of maize: Chaparreño.

This is the only jar I have seen that depicts a frog, which is symbolically associated with rain. It is in the typical red and white Moche style. The body of the redware olla is shaped like a frog with ears, eyes, and snout. There is an ear of maize on each side of the body and manioc tubers around the rear of the animal. The plain spout, maize, manioc, and parts of the face are painted white.

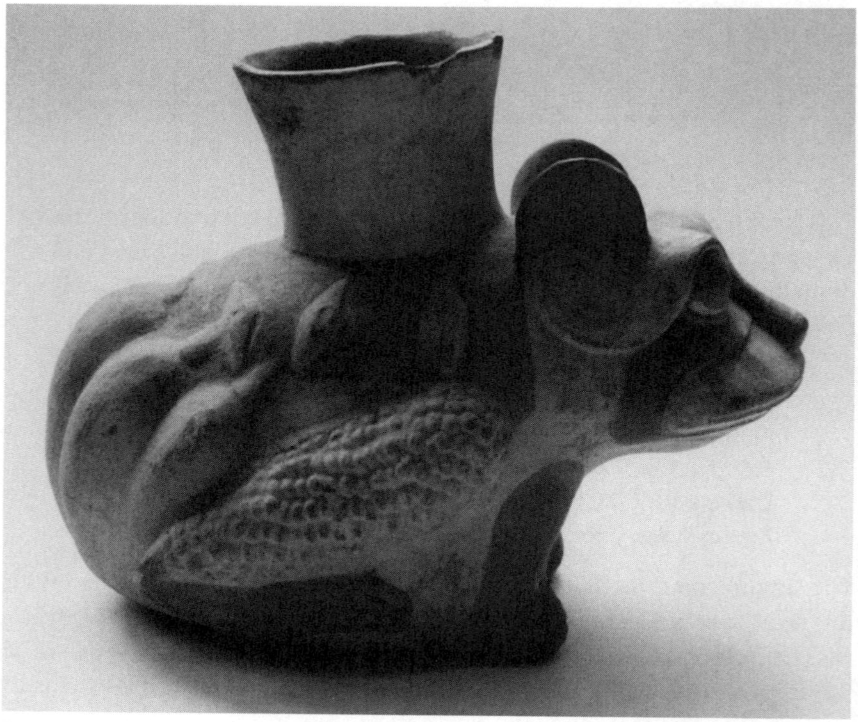

Fig. 66. P-56. Olla depicting a frog with maize and manioc.

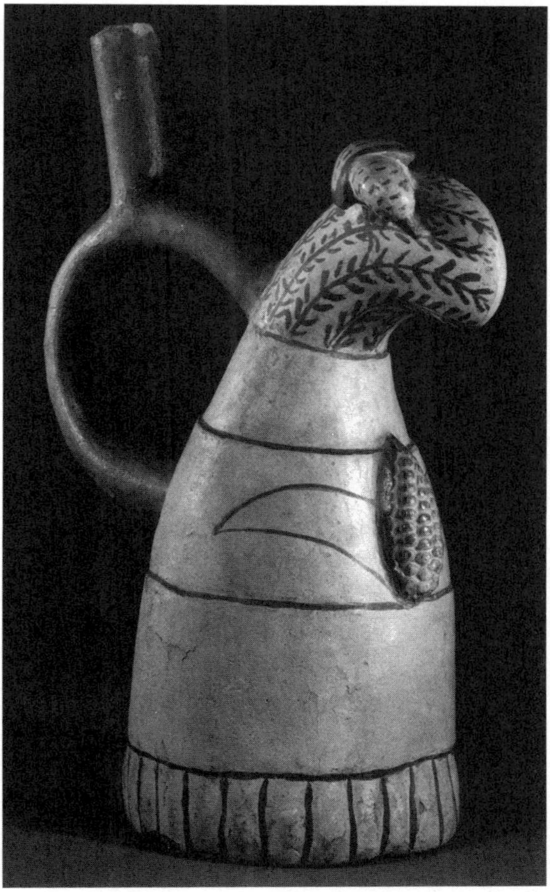

Fig. 67. P-57. Stirrup-spout jar depicting a maize plant.

Stirrup-Spout Jar Depicting a Maize Plant

P-57. Fig. 67. Stirrup-spout jar depicting a maize plant.
Provenience: excavated in the Moche Valley.
Metropolitan Museum of Art 67.167.19.
Fabric: red and white.
Dimensions: ht. 28.1 cm, width 20.6 cm.
Race of maize: Chapalote/Nal Tel.

This is a Moche IVA stirrup-spout jar which represents a maize plant with a bird perched in the tassel. The redware vessel is covered with a cream or white slip and is divided into five sections marked by red lines painted around the jar. The bottom section is marked by red vertical stripes. Above this is a plain section. The midsection has a mold-made ear of maize attached to the jar in the front of the vessel and one in the back. Leaves or husks of the ears are painted on the jar, spreading out from the ears. The section above the midsection is undecorated and is surmounted by the top section, which curves around and has tassels painted on it. The bird is a parrot (*lorito*) and is perched on the top of the jar leaning toward the front. Its markings, the curved beak, and eyes are painted red. Lorito birds are frequently depicted in association with corn on Nazca and North Coast art (Grobman et al. 1961). Loritos are still one of the worst pests of corn in many parts of Peru.

Stirrup-Spout Jars of Mounds of Maize

P-58. Fig. 68. Stirrup-spout jar depicting a mound of maize.
Provenience: excavated in the Virú Valley.
Museo Nacional de Antropología y Arqueología 1/1681.
Fabric: red.
Dimensions: ht. 20.8 cm, width 15.0 cm.
Race of maize: Chapalote/Nal Tel.

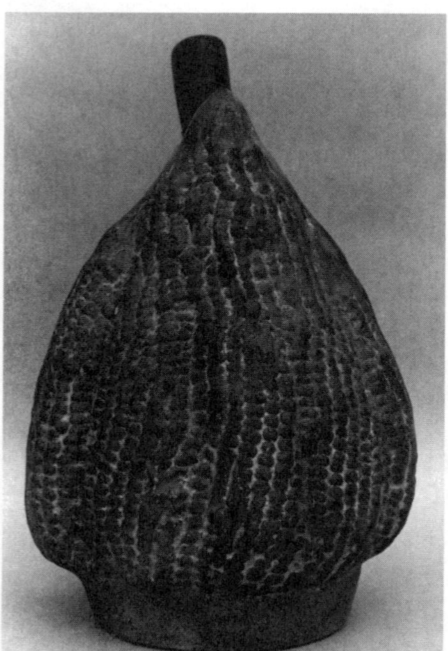

Fig. 68. P-58. Stirrup-spout jar depicting a mound of maize.

This Moche IVA stirrup-spout jar is covered with maize as if to represent a mound of maize. The maize ears are so well blended into the body of the jar that they are barely distinguishable. The tips of the maize ears have been blended together to form one tip at the top of the jar.

P-59. Stirrup-spout jar depicting a mound of maize.
Provenience: excavated in the Virú Valley.
Museo Nacional de Antropología y Arqueología 1/1680.
Fabric: red and white.
Dimensions: ht. 19.7 cm, width 18.0 cm.
Race of maize: Confite Puneño/Quicheño Rojo.

The style of this jar resembles the one described for P-58.

P-60. Stirrup-spout jar depicting a mound of maize.
Provenience: excavated in the Virú Valley.
Museo Rafael Larco Herrera 4677.
Fabric: red.
Race of maize: Güirua.

Barren tips are represented on the maize ears, and the jar terminates in a barren tip.

P-61. Fig. 69. Stirrup-spout jar depicting a mound of maize.
Provenience: unknown.
Museo Rafael Larco Herrera S-39160.

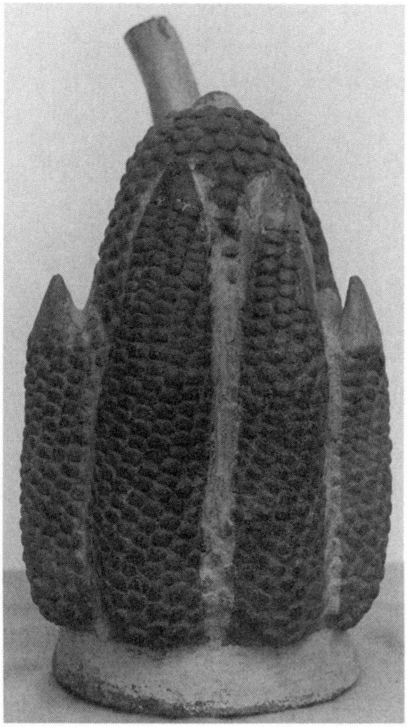

Fig. 69. P-61. Stirrup-spout jar depicting a mound of maize.

Fabric: red and white.
Race of maize: Mochero.

This Moche IVA redware jar depicts a mound of maize in a somewhat different fashion. There are four mold-made ears on the front of the jar. The main body of the jar, however, is covered with a stylized kernel pattern and is shaped like the other jars representing a mound of maize. The stirrup spout and the annular base are painted white. White also outlines the four ears of maize. The maize ears have barren tips.

P-62. Fig. 70. Stirrup-spout jar depicting a mound of maize.
Provenience: Chimbote.
Museum für Völkerkunde Berlin 17836.
Fabric: red and white.
Dimensions: ht. 18.1 cm, width 12.5 cm.
Race of maize: Pira/Nal Tel.

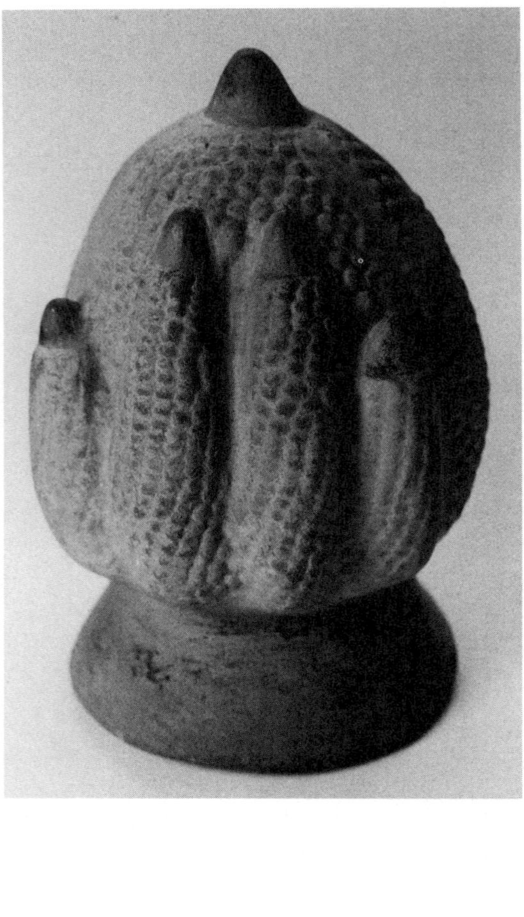

Fig. 70. P-62. Stirrup-spout jar depicting a mound of maize.

Although the stirrup-spout is broken, it is evident where it was attached, and this jar resembles others that depict a mound of maize. All of the maize ears have barren tips. The spout, base of the jar, and tips of the ears are highly polished redware; the maize is covered with a white slip.

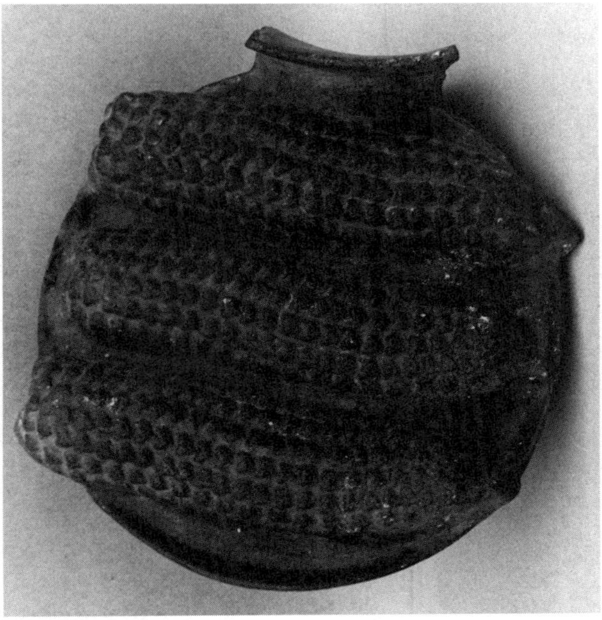

Fig. 71. P-63. Bowl of maize.

Bowl of Maize

P-63. Fig. 71. Bowl of maize.
Provenience: excavated in the Virú Valley.
Museo Nacional de Antropología y Arqueología 1/2609.
Fabric: red.
Dimensions: ht. 10.7 cm, width 17.0 cm.
Race of maize: Imbricado.

This object is unusual because it is not a functional container. It represents a bowl of three maize ears. It is redware and is definitely a Moche period piece. This is the only specimen I have seen that could not serve as a container of some kind. It is notable because it depicts distinctly imbricated kernels.

Ollas Representing Baskets of Maize

P-64. Fig. 72. Olla depicting a basket of maize.
Provenience: unknown.
Museo Arqueológico Brüning de Lambayeque 7761.
Fabric: black.
Race of maize: Confite Morocho.

This blackware vessel is decorated with ears of maize and fish designs in low relief. All of the vessels in this group are distinguished by their composite silhouette bodies. The constricted neck of this vessel has a flaring rim.

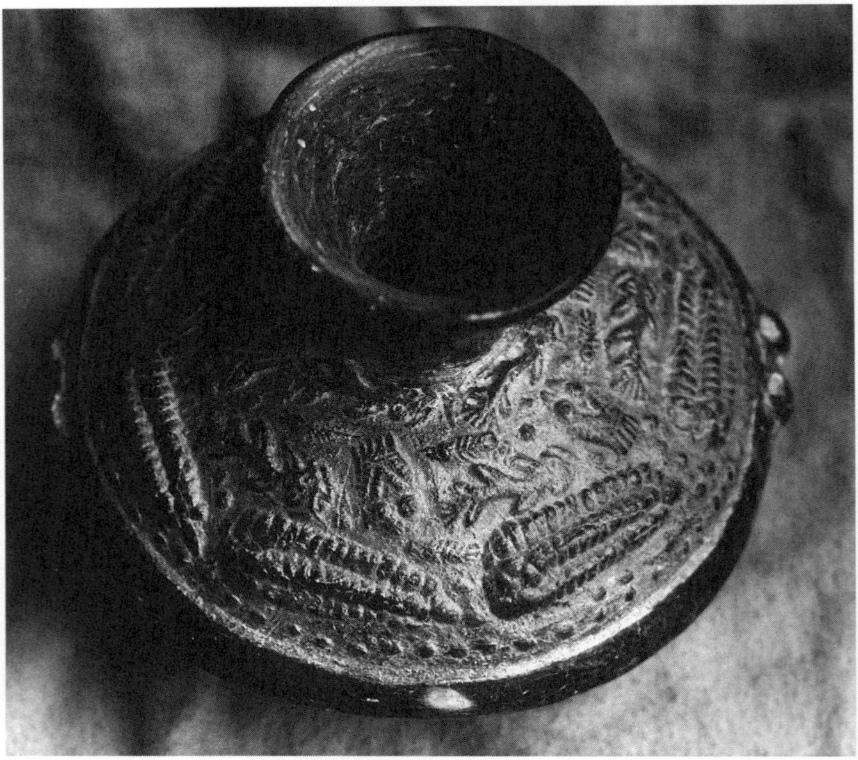

Fig. 72. P-64. Olla depicting a basket of maize.

P-65. Fig. 73. Olla depicting a basket of maize.
Provenience: unknown.
Museo Arqueológico Brüning de Lambayeque B-1918.

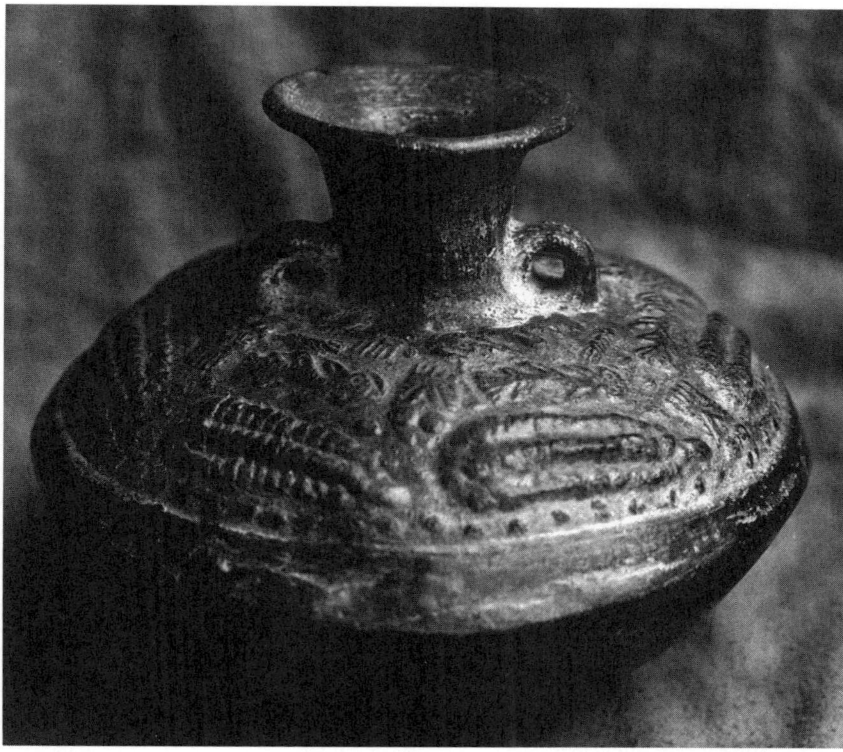

Fig. 73. P-65. Olla depicting a basket of maize.

Fabric: black.
Race of maize: Confite Morocho.

This vessel is virtually identical to the one described for P-64. The noticeable difference is the presence of two small loops on either side of the spout, which may have been used for threading a cord for carrying.

P-66. Fig. 74. Olla depicting a basket of maize.
Provenience: unknown.
Museo Arqueológico Brüning de Lambayeque B-1917.
Fabric: black.
Race of maize: Pollo.

This vessel's flattened shape gives it a basket-like appearance. Fourteen ears of maize adorn the top of the olla. The spout is slightly flaring and is flanked by small loops, which may have been used for threading a carrying cord.

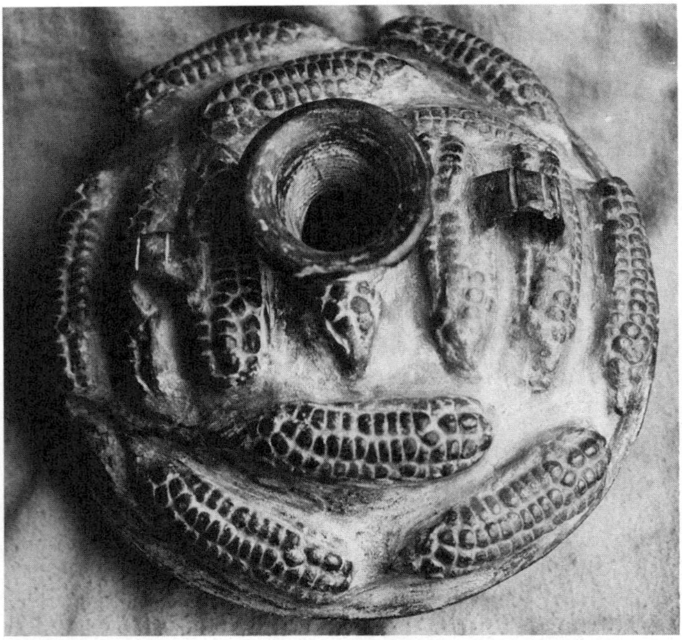

Fig. 74. P-66. Olla depicting a basket of maize.

P-67. Fig. 75. Olla depicting a basket of maize.
Provenience: unknown.
Museo Arqueológico Brüning de Lambayeque MB-1927.
Fabric: red and black.
Race of maize: Karapampa.

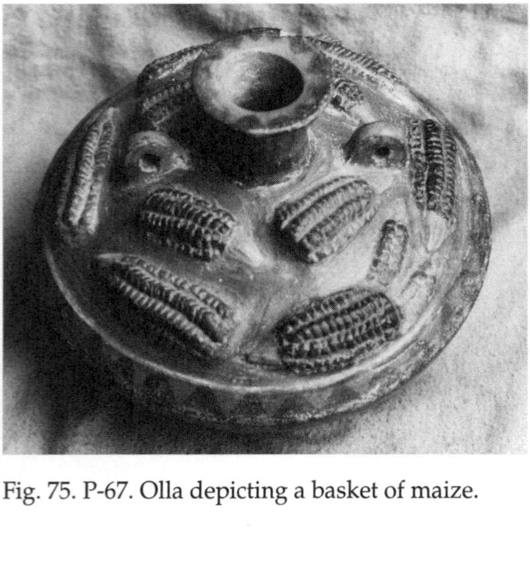

Fig. 75. P-67. Olla depicting a basket of maize.

The edge of the everted rim of the jar and the rim of the upper half of the vessel body are decorated with a black zigzag design. The maize on this jar is also painted, some ears red and others black. All the ears have barren tips. This jar also has small loops presumably for threading a carrying cord.

P-68. Fig. 76. Olla depicting a basket of maize.
Provenience: unknown.
Museo Nacional de Antropología y Arqueología 36/1376.
Fabric: red and white.
Dimensions: ht. 15.0 cm, diameter 21.0 cm.
Race of maize: Karapampa.

The distinguishing difference between this jar and the one described for P-67 is that the zigzag decoration on P-68 is painted white and is only on the outer rim of the vessel. In addition, the ears of maize are not painted.

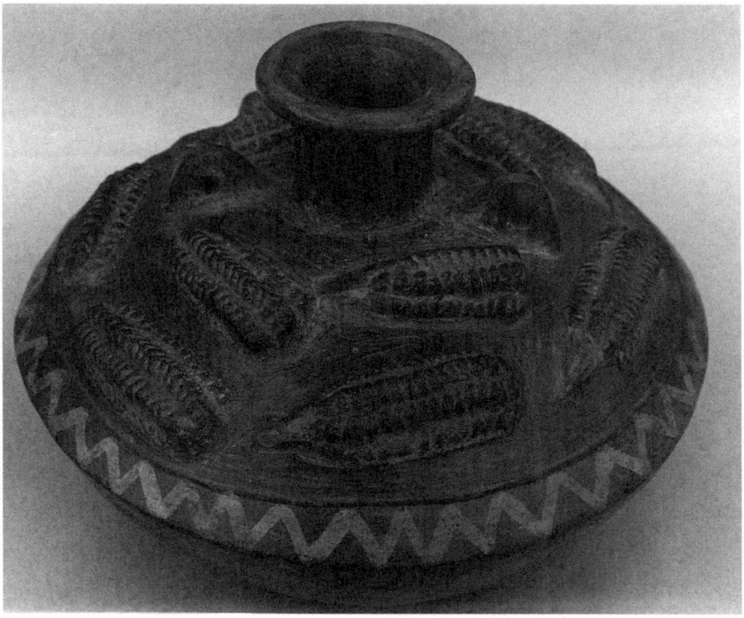

Fig. 76. P-68. Olla depicting a basket of maize.

P-69. Fig. 77. Olla depicting a basket of maize.
Provenience: unknown.
Duke University Museum of Art 73.1.408.
Fabric: red and dark brownish-red.
Dimensions: ht. 10.6 cm, diameter 17.5 cm.
Race of maize: Pira/Nal Tel.

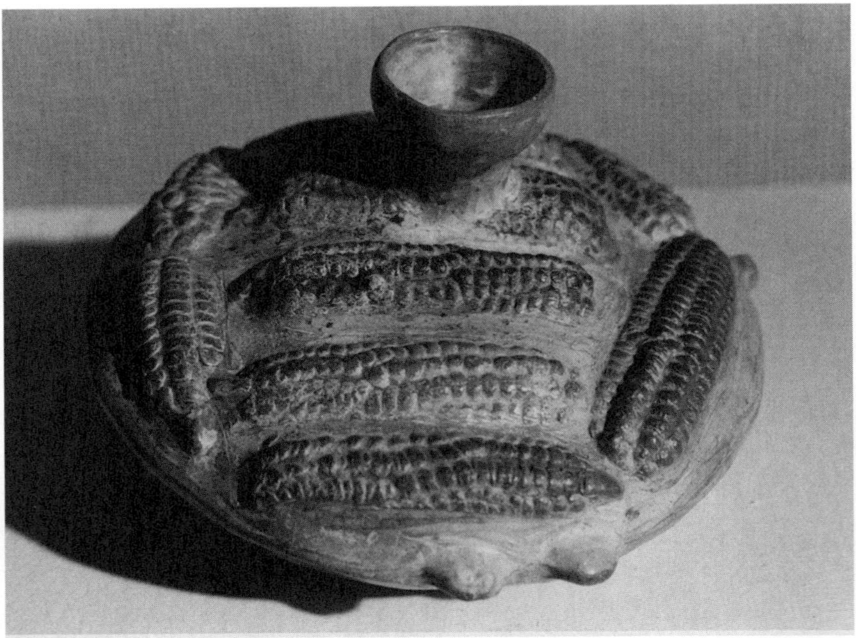

Fig. 77. P-69. Olla depicting a basket of maize.

This vessel may have been used for containing chicha, an alcoholic beverage made from maize. Chicha was an important ceremonial drink, providing a means of intercession with the gods (Nicholson 1960, 290). The small orifice on the side of the vessel would have been a convenient pouring spout that could be plugged until time for consumption. The ears of maize on this jar are painted dark reddish brown and dark red, colors typical of maize used in making chicha.

P-70. Fig. 78. Olla depicting a basket of maize.
Provenience: Trujillo.
Museum für Völkerkunde Berlin 3571.
Fabric: black.
Dimensions: ht. 13.8 cm, width 18.8 cm.
Race of maize: Pira/Nal Tel.

Although the blackware is indicative of the later Chimú culture, the shape of the olla and ears of maize on it resemble P-68 and P-69. The ears of maize are depicted with barren tips.

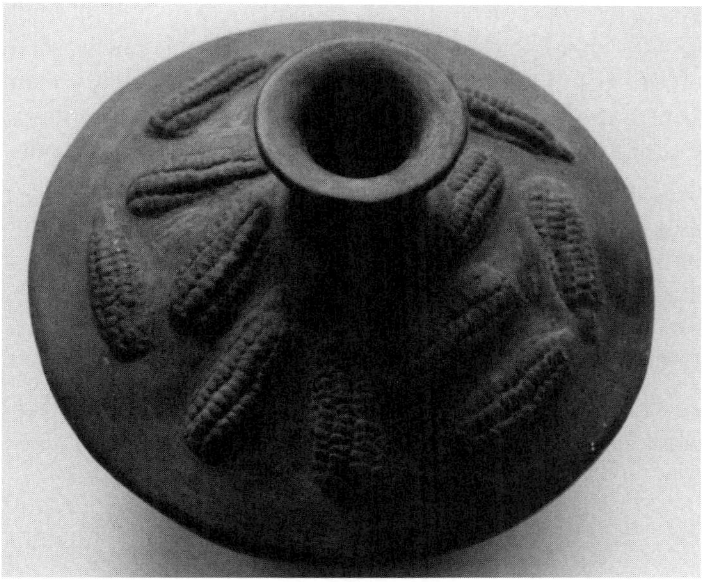

Fig. 78. P-70. Olla depicting a basket of maize.

P-71. Fig. 79. Olla depicting a basket of maize.
Provenience: unknown.
Museum of the American Indian, Heye Foundation 7/2734.
Fabric: black.
Dimensions: ht. 16.3 cm, diameter 12.5 cm.
Race of maize: Pira/Nal Tel.

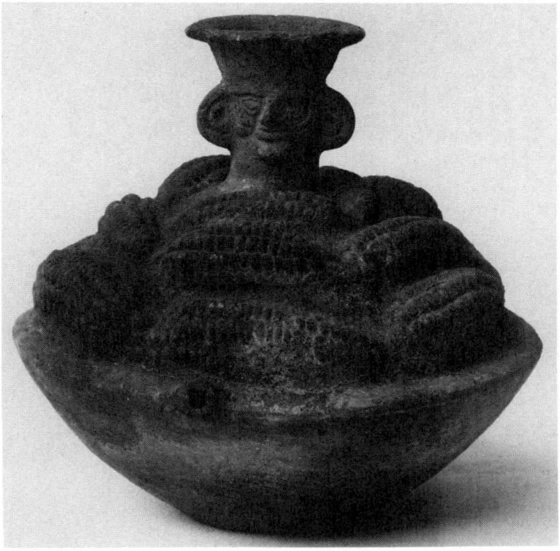

Fig. 79. P-71. Olla depicting a basket of maize.

This vessel may also have been used for containing chicha. It has the same small orifice on the side of the jar as does P-69. There is the face of a man, without fangs, on the neck of the spout. This is a decorative style diagnostic of the Chimú culture. Moche jars characteristically have unadorned spouts.

P-72. Fig. 80. Olla depicting a basket of maize.
Provenience: unknown.
British Museum 1921.10.27.
Fabric: black.
Dimensions: ht. 17.2 cm, diameter 23.8 cm.
Race of maize: Pira/Nal Tel.

This blackware Chimú jar has a narrow, flanged neck that is partially broken. Thirteen mold-made maize ears adorn the upper flared section of the vessel.

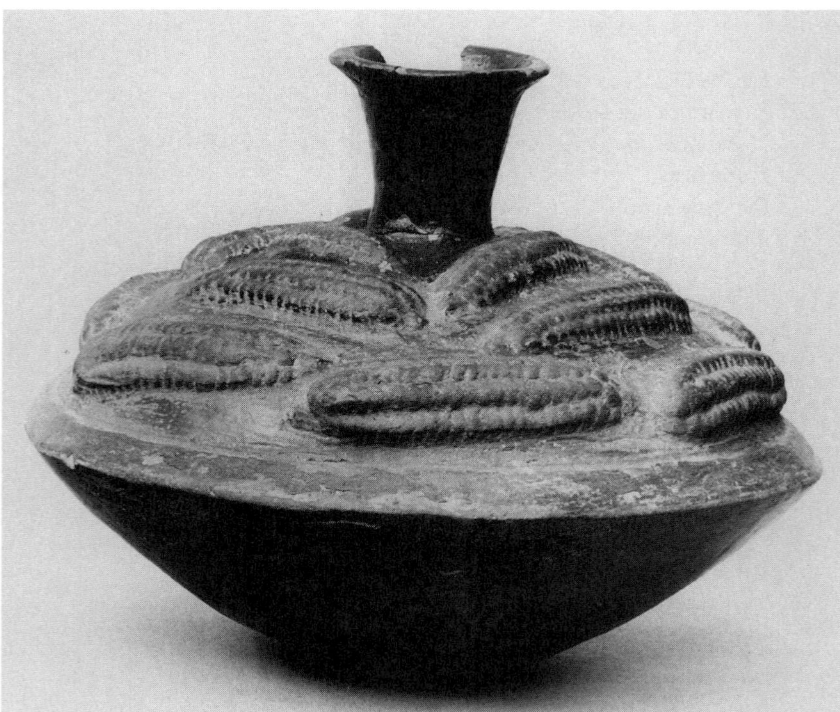

Fig. 80. P-72. Olla depicting a basket of maize.

3

Zapotec Urns

Oaxaca: Land of the Zapotec

The Valley of Oaxaca is in the highlands of southern Mexico. The capital of the state of Oaxaca is Oaxaca City, located at the convergence of three arms forming a Y-shaped configuration that includes the Etla Valley to the west, the Tlacolula (or Mitla) Valley to the east, and the Zaachila Valley to the south (fig. 81). The average elevation of the valley floor is 1,550 meters. The climate is semiarid with an annual rainfall of 500–700 millimeters that occurs primarily during the summer months (Flannery et al. 1967, 445–49). Oaxaca is so dry that it is classified by geographers as an area of permanent drought (Marcus and Flannery 1996). Although the temperatures in the valley may be well adapted to year-round planting of maize, water is scarce and maize is planted seasonally to correspond with the approaching rainy season from April to June (Kirkby 1973, 54–55). The Zapotec Indians inhabited this area since around 60 B.C.

Archaeological Context

Monte Albán, located on a mountain just outside the city of Oaxaca, was the Zapotec political and religious capital for centuries. Although the urns portrayed in this study come from all parts of the valley, most excavated pieces are from Monte Albán. The archaeological record for the Valley of Oaxaca demonstrates cultural continuity from as early as Monte Albán I until the present day (Marcus and Flannery 1996). On the basis of a radiocarbon date from Monte Negro in association with a Monte Albán I level, this period begins as early as 650 B.C. (Rabin 1970, 14). The archaeological cultural remains have been attributed to the Zapotecs (Jiménez M. 1970, 12), and no scholar has seriously questioned this dating.

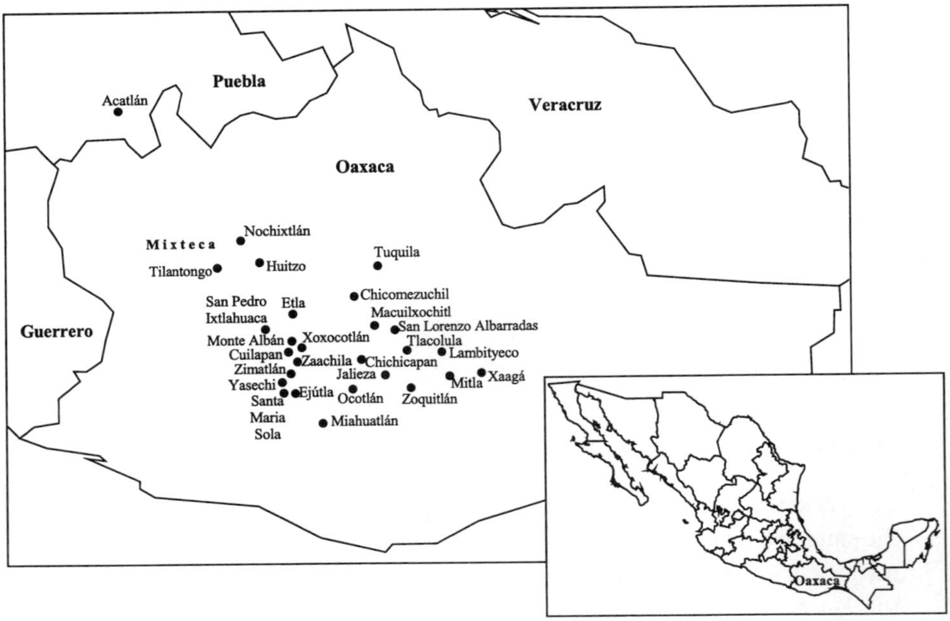

Fig. 81. Map of Oaxaca showing proveniences for urns depicting mold-made maize.

The Valley of Oaxaca has a long history of human occupation extending back to the late Pleistocene around 15,000 B.C. or earlier (Marcus and Flannery 1996). The first humans in the valley were bands of hunters and gatherers. About 5000 B.C., the Indians had made the transition to incipient cultivators growing beans, maize, and squash in forest clearings. Through time, with the development of agriculture and an expanding human population, complex social organization evolved and gave rise to the state and to Zapotec civilization with the large urban center of Monte Albán. The periods Monte Albán I–V were established on the basis of ceramic styles from the excavations at Monte Albán. The corresponding chronology for Monte Albán I is 500–100 B.C.; Monte Albán II, 100 B.C.–A.D. 200; Monte Albán III, A.D. 200–750; Monte Albán IV, A.D. 750–1000; Monte Albán V, A.D. 1000–1300 (Marcus and Flannery 1996). The urns that are the subject of this study are assignable to Monte Albán III and IV, which occured during the Mesoamerican Classic and Early Postclassic periods, respectively.

Monte Albán III is divided into two phases, IIIA circa A.D. 200–500 and IIIB circa A.D. 500–750 (Marcus and Flannery 1996). There was dramatic growth in the Valley of Oaxaca during the Early Classic period. The number of sites doubled in Monte Albán IIIA to a thousand with a total population estimated at 115,000. Monte Albán during IIIB was marked by extensive building activity and continued population growth in Oaxaca (Kowalewski et al. 1989). On the basis of the use of the mold and the association of certain deities with maize, both features attributed to Monte Albán IIIB (Shaplin 1975; Paddock 1970, 193), the urns with depictions of mold-made maize are presumed to belong to IIIB. Some urns can be assigned to the Monte Albán IIIB–IV transition from Late Classic to Early Postclassic, a time when the trend was away from centralized government to greater regionalization with more emphasis on market activity and craft production at other sites. During the height of Monte Albán influence and control in IIIA and IIIB, the main centers were located in the Valle Grande around the capital city of Monte Albán and in the Etla arm to the north. During Monte Albán IV there was a shift to greater occupation of the sites in the Tlacolula arm of the valley. The urns with mold-made maize are either Monte Albán IIIB (roughly A.D. 500 to 750) or Monte Albán IV (roughly A.D. 700 to 1000) (Marcus and Flannery 1996, 25). The ceramics of these two periods are often difficult to differentiate and may be assigned to both. The city of Monte Albán probably reached its peak during Monte Albán IIIB, when the population is estimated at 24,000 (Marcus and Flannery 1996, 234). After IIIB, the city began to decline as the population dispersed to other centers in the valley during Monte Albán IV. Jalieza was the largest cemetery during Monte Albán IV. During this five-hundred-year span, the population of the Valley of Oaxaca is estimated to have been greater than 100,000.

Zapotec Urns

Zapotec urns have been found in antechambers and interiors of tombs (Caso and Bernal 1952, 10). They were placed in niches on outside walls, above entrances, and on roofs of tombs (Saville 1899, 361; Saville 1904, 51) and have been found as part of offerings buried in the floors of ceremonial centers (Lothrop 1972, 76). The urns maintained their form for a millennium and are typical of Zapotec culture (Anton 1969, 48). Some authors suggest that the urns contained food offerings to the dead, were used as incense burners, or contained ashes of the dead (Linné 1938, 90; Mason 1929, 177; Saville 1904, 60). Nevertheless, the urns have practically always been found empty (Caso and Bernal 1952, 10; Anton 1969, 48). On occasion, offerings of obsidian blades, jade beads, snail shells used as bells, or bones of small animals have been found inside urns (Caso and Bernal 1952, 10). In addition, Caso reported the find of an urn containing some greenstone figurines of Teotihuacán style and some jade beads in a Monte Albán IIIA offering buried in Monticulo I at Monte Albán (1938, 10). Zapotec urns are not funerary urns in a strict sense, but their primary function was as grave goods. Although outside influences can be noted, artistically these urns "bear the unmistakable stamp of the Zapotec artist's own peculiar psychology" (Joyce 1927, 99). There is no other art form like these urns in Mesoamerica, and certain notable artistic conventions recur in the style. Most urns are found at important population centers engaged in civic-ceremonial activities (Kuttruff 1978). Most have been found at the top of Monte Albán, diminishing in frequency as distance from the Main Plaza increases. The next greatest frequency of occurrence is at secondary centers throughout the valley. Some are found in towns, and they are rarely found in villages (Kowalewski et al. 1989). Of the urns in this study for which proveniences are available, twenty-two come from the Valle Grande and one from Etla. Three come from the Tlacolula arm, two are from the Mixteca, and four are from other areas. These are listed in table 3.1. When mazorca fragments for which proveniences are recorded are included in a survey of sites where naturalistic maize depictions are found, the number increases to thirty-eight in the Valle Grande, fourteen in the Etla Valley, twelve in the Tlacolula Valley, three from the Mixteca, and nine from other areas. The Valle Grande sites are Monte Albán, Xoxocotlán, Cuilapan, Zaachila, Zimatlán, Yasechi, Jalieza, Ocotlán, Ejútla, and Miahuatlán; the Etla Valley sites are Etla, San Pedro Ixtlahuaca, and Huitzo; the Tlacolula Valley sites are Macuilxóchitl, Tlacolula, Lambityeco, Mitla, and Xaagá; other locations include the Mixteca, Nochixtlán, Tilantongo, Tuquila, Chicomezuchil, Zoquitlán, Chichicapan, San Lorenzo Albarradas, Acatlán in Puebla, Chalco, and Morelos.

Description

The ceramic urns are made of a highly fired, porous gray ware. They are usually composed of a cylindrical vessel with an anthropomorphic or theriomorphic figure built out in front of the cylinder. Most authorities think the figures represent deities from the Zapotec pantheon. Joyce Marcus (1983, 144–48), however, asserts that rather than representing divinities, the figures on Zapotec urns may depict deceased ancestors or personages wearing masks representing figures with supernatural attributes. For the sake of consistency and ease of cross-referencing, the conventional scheme of describing the categories as deities is followed here. I must emphasize, however, that these are anthropomorphic figures associated with

Table 3.1. Known proveniences of Zapotec urns with mold-made maize depictions

Specimen	Provenience
Valle Grande	
MNA 6-6223 (Z-1, Figs. 82, 83)	Tomb 124, Monte Albán
MNA 6-2293 (Z-2, Fig. 84)	Tomb 40, Monte Albán
MNA 6-2290 (Z-3)	Tomb 40, Monte Albán
MNA 6-2289 (Z-4)	Tomb 40, Monte Albán
MNA 6-6717 (Z-5, Fig. 85)	Patio 1 of Tomb 138, Monte Albán
MNA 6-6754 (Z-6, Fig. 86)	Antechamber of Tomb 31, Monte Albán
MARI 35-5976 (Z-54, Fig. 123)	Monte Albán
AMNH 30/6335 (Z-37)	Xoxocotlán, mound 9, tomb 3 façade
AMNH 30/6334 (Z-36, Fig. 108)	Xoxocotlán, mound 9, tomb 3 façade
Berlin 11412	Cuilapan
MNA 6-338 (Z-8, Fig. 88)	Zaachila
MNA 6-6757 (Z-13, Fig. 92)	Zaachila
MNA 6-328 (Z-18)	Zaachila
MNA 6-6672 (Z-31, Fig. 106)	Zaachila
Frissell 4279 (Z-34, Fig. 107)	Zaachila
Paris 24.13.3645 (Z-41, Fig. 111)	Zaachila
MNA 6-6894 (Z-46, Fig. 116)	Zaachila
AMNH 30/11173 (Z-39, Fig. 109)	Zimatlán
Berlin 28354 (Z-49, Fig. 119)	Zimatlán
AMNH 30/95 (Z-38)	Yasechi (Santa Inés Yatzeche)
Frissell 4889 (Z-43, Fig. 113)	Ejútla
MNA 6-5421 (Z-12, Fig. 91)	Miahuatlán
Etla	
MNA 6-6885 (Z-45, Fig. 115)	Etla
MNA 6-965 (Z-47, Fig. 117)	San Pedro Ixtlahuaca
Berlin 29029 (Z-48, Fig. 118)	San Pedro Ixtlahuaca
Tlacolula	
Berlin 31339	Tlacolula
Frissell 6883 (Z-7, Fig. 87)	Lambityeco (excavated)
MARI 35-6004 (Z-55, Fig. 124)	Mitla
Mixteca	
MNA 6-6708 (Z-28, Figs. 102, 103)	The Mixteca
MNA 6-6758 (Z-29, Fig. 104)	Nochixtlán
Other	
Munich 14.8.6 (Z-22)	Tuquila
NMNH 115033 (Z-52, Fig. 121)	Chicomezuchil
Senter U5/4552 (Z-56, Fig. 125)	Morelos
NMNH 115032 (Z-9, Fig. 89)	Acatlán (Puebla)

attributes assigned to certain deities, attributes that are symbolic of aspects of Zapotec life such as the calendar, life and death, sustenance, and everyday activities (Boos 1966a, 19), and we do not know what they were meant to symbolize.

The figures are most frequently depicted seated cross-legged and wearing elaborate headdresses. A relative degree of importance is assigned to different parts of the figure, shown by the size and attention to detail given them. These include, beginning with the most important element and following in degree of importance, the headdress, technically termed the *tocado*, the most important and outstanding feature of an urn. The elements of the headdress have special significance for the divinity or its *nagual*, the guardian spirit or alter ego of the god the urn may represent (Boos 1966a, 17). The second most important aspect is the face, which often has a characteristic nasal or buccal mask. Third is the pectoral ornament, a necklace that frequently has a characteristic glyph medallion displayed on the figure's chest. Next to be considered is the figure's personal jewelry, which indicates status. Last come the arms, hands, legs, and feet. The limbs are often clumsily modeled and are large, mold-made, and attached at odd angles. The feet are usually out of proportion, roughly modeled, with two to four toes. The posture most common for these figures is a cross-legged seated position with the palms of the hands placed on the knees.

Classification of the deities represented on Zapotec urns follows Alfonso Caso and Ignacio Bernal's (1952) 44 main categories and 138 subcategories. Table 3.2 lists characteristic attributes of eight Zapotec deities and an unidentified figure associated with naturalistic maize.

Of all categories, urns representing Cocijo are most numerous (Caso and Bernal 1952, 17). Cocijo, God of Lightning and Rain, was the most important deity in the Zapotec pantheon. Since maize was the mainstay of pre-Columbian life and the development and survival of urban society depended on its successful production, this divinity, who controlled the elements on which the success of maize and other crops depended, was of paramount importance. Cocijo, however, is not included in Caso and Bernal's "Maize Complex," and of the many Cocijo urns in existence, comparatively few depict maize. The primary attribute associated with Cocijo is the face mask with its characteristic elements: the bifurcated tongue protruding from open jaws representative of a serpent; the nasal and buccal elements, which are bands across the nose and cheeks representative of a feline; "merlon-shaped eyebrows" said to represent the heavens, a high place, or infinity; and the squared, indented, lower eyelids representing clouds (Boos 1966a, 27). Although there are exceptions where it is

Table 3.2. Characteristic attributes of deities on Zapotec urns depicting maize

Deity	Tocado	Face	Attire	Posture	Race of maize
God of Glyph "L" (Pitao Cozobi), the principal maize deity	"C" glyph forms central medallion	Aspas, nasal mask, buccal mask	Pectoral glyph, loincloth, earspools	Seated cross-legged	Nal Tel/Pira, Karapampa, Chapalote, Mochero
Cocijo, God of lightning and rain	"C" glyph forms central medallion (exceptions where not present)	Nasal mask, buccal mask, "merlon" eyebrows, squared lower eyelids, extended forked tongue	Pectoral glyph, loincloth, earspools	Seated cross-legged	Nal Tel/Pira, Confite Puntiagudo
God with the ear of maize in his headdress	"C" glyph superimposed on ear of maize forms central medallion; or figure with two hands holding an ear of maize as central ornament; or entire tocado consists of standing ears of maize	Unmasked	Pectoral glyph, loincloth, earspools	Seated cross-legged or sometimes standing	Nal Tel
God displaying the mask of the God Cocijo as	Cocijo mask as central medallion; pair of hands	Unmasked youthful personage	Pectoral glyph, loincloth, earspools	Seated cross-legged	Unidentified

(continued)

Table 3.2.—*continued*

Deity	Tocado	Face	Attire	Posture	Race of maize
the medallion of the headdress	suspended in lower part; trifoliate element and maize				
God with the bowknot in his headdress	Bowknot as central medallion	Unmasked or wearing mask of God of Glyph "L" or Cocijo	Pectoral glyph, loincloth, ear-spools, bead necklace	Seated cross-legged	Chapalote/Nal Tel
Goddess "13 Serpent"	Woven or plaited strands of hair, often adorned with ball-shaped elements	Unmasked	Huipil, quichquemitl, bead necklace, earspools	Seated cross-legged	Oke
Opossum God		Round, staring eyes; nose with indented ridge spiral up to forehead; lips pulled back baring teeth	Pectoral glyph, loincloth	Seated cross-legged or standing	Nal Tel/Pira
Goddess "2 J"	"J" glyph forms central medallion	Unmasked	Huipil, quichquemitl, bead necklace, earspools	Generally appears standing or kneeling	Dent corn
Unidentified					Nal Tel/Pira

not displayed, the "C" glyph, centrally located in the headdress, is an important attribute.

The God of Glyph "L" was so named by archaeologists because of the elements above and below each eye which resemble the Zapotec Day-Sign referring to movement or earthquakes. This Day-Sign was labeled glyph "L" (Caso 1928, 39–40), and by extension Pitao Cozobi, which is the Zapotec name for this deity, who is more popularly known in the literature as the God of Glyph "L," although no "L" glyph is actually represented with the figure, and the deity is not associated with earthquakes. Characteristic attributes of the divinity are the eye elements, referred to as *aspas,* a nasal mask in the form of a nose knot, a buccal mask in the form of lines curving around each cheek, and the "C" glyph forming the central attribute of the headdress. This god first appears during Monte Albán IIIA. Thereafter, it became quite widespread (Boos 1966a, 177–78).

The category of the "God with the Ear of Maize in His Headdress" is generally represented by a figure with a "C" glyph as the central medallion of the headdress, which is superimposed by a stylized ear of maize. For the purpose of this work, this category has been extended, as proposed by Frank Boos (1966b, 348), to include figures that have two hands holding a naturalistic ear of maize as the central attribute of the headdress and some that have naturalistic ears of maize standing vertically in the headdress.

The "God Displaying the Mask of the God Cocijo as the Medallion of the Headdress" is another deity. A diagnostic characteristic of this category is that all the figures are young, sometimes adolescent. The headdress is centered around the god Cocijo, depicted with the characteristic facial mask and "C" glyph and having a pair of hands suspended in the lower part of the headdress and on each side of the mask. A few of the figures have a second pair of hands in the headdress. The face of the personage on the urn is typically represented without any mask. A trifoliate element and maize symbol, also characteristic of this category, occurs on each side of the Cocijo mask in the headdress (Boos 1966a, 65). It represents a stylized staminate tip and silks of an ear of maize. The staminate tip is a spike of male flowers borne directly above the female flowers that form the seed upon pollination. Normally, the male flowers of maize and its closest relative, a wild Mexican grass called teosinte, are borne separately on the plant in a tassel at the apex of the culm (stalk). The staminate tip is a primitive botanical trait that occasionally still appears in certain ancient indigenous maize races, Nal Tel and Chapalote of Mexico, Pollo of Colombia, and Confite Morocho of Peru (Mangelsdorf 1958, 460). This feature is typical of the wild grass *Tripsacum,* a more distant relative of maize.

Tripsacum has been postulated to be one of the progenitors of maize (Mangelsdorf and Reeves 1939), and recent evidence lends support to this idea (Eubanks 1995, 1997). The consistent stylized depiction of the staminate tip embellishing molded maize ears on Zapotec urns underscores that this tripsacoid trait had symbolic significance for the Zapotec. It may reflect Zapotec knowledge and ritual recognition of this important source of ancestral germplasm in ancient maize races.

The "God with the Bowknot in His Headdress" is another deity found in connection with depictions of maize. The main diagnostic of this category is the bowknot as the central ornament of the headdress. The personage may appear unmasked or wearing a mask characteristic of Cocijo or the God of Glyph "L" (Boos 1966a, 352). Of special interest are two Moche jars depicting a similar bowknot motif in the headdress (cf. fig. 61), an unusual feature that may have been adapted as a result of contact between the Zapotec and Moche cultures.

The Goddess "13 Serpent" is characterized by a headdress of woven strands often adorned with ball-shaped ornaments that may represent bells (Boos 1966a, 352).

Another deity represented is the Opossum God, whose diagnostic elements include round, staring eyes, an indented nose ridge that forms a spiral extending to the forehead, and lips pulled back showing the teeth (Boos 1966a, 237).

The Goddess "2 J" is named for the "J" glyph depicted in the headdress. She wears the indigenous feminine dress, the *huipil* or undergarment, and the *quichquemitl*, the bordered outer garment. She generally is portrayed standing with her arms crossed on her bosom or hanging by her sides (Boos 1966b, 338).

In addition to these deities, there are three vessels encircled by ears of maize. The Goddess "2 J" appears on one of these vessels, and the others are without figures.

Fifty-four Oaxacan vessels are presented in the following order: urns depicting the God of Glyph "L," urns depicting Cocijo, urns depicting the God Displaying the Mask of the God Cocijo as the Medallion of the Headdress, urns depicting the God with the Ear of Maize in His Headdress, urns depicting the God with the Bowknot in His Headdress, urns depicting the Goddess "13 Serpent," urns depicting the Opossum God, and vessels encircled with ears of maize. The pieces are introduced in catalog format. Each is designated by a number preceded by the letter Z. After the catalog number, basic information, including the figure in which the specimen is illustrated, vessel type, provenience, museum location and museum catalog number, fabric, dimensions, chronological period to which the piece has been assigned, and race of maize identified on the specimen are given. A description of each piece follows the catalog entry information.

Urns depicting the God of Glyph "L"

Z-1. Figs. 82 and 83. Urn of the God of Glyph "L."
Museo Nacional de Antropología 6–6223.
Provenience: Tomb 124, Monte Albán.
Fabric: dark gray.
Dimensions: ht. 34.0 cm, width 29.0 cm.
Period: Monte Albán IIIB.
Race of maize: Karapampa.

This piece is assigned to Monte Albán III-B because the form of the "C" glyph which is the central medallion of the headdress was becoming extinct during Monte Albán IV (Leigh 1970, 263). John Paddock, however, said that the row of four tassel elements bordering the loincloth did not come into use until Monte Albán IV (personal communication), so there is

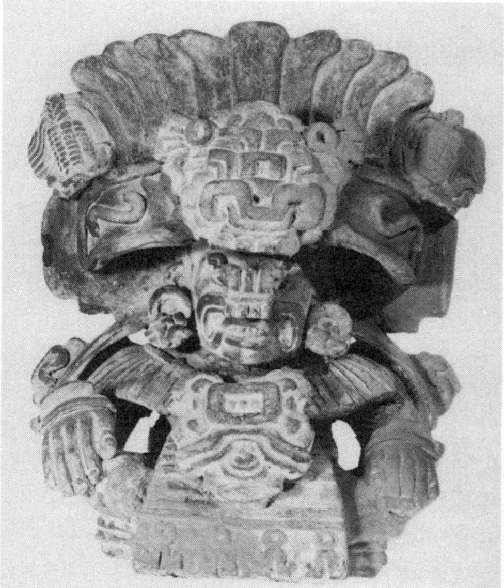

Fig. 82. Z-1. Urn of the God of Glyph "L."

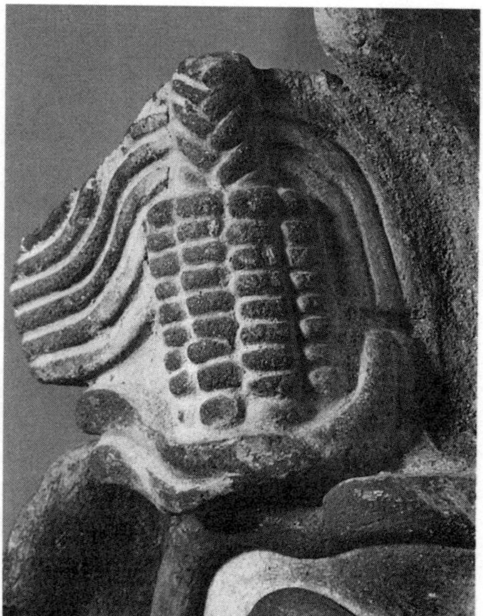

Fig. 83. Z-1. Closeup of maize ear in fig. 82.

ambiguity in the chronological assignment. The headdress on this figure is quite elaborate, the central medallion being surmounted by elements representing feathers. This feather adornment in the headdress is referred to as a *penache*, and here it is flanked on either side by an ear of maize displaying a staminate tip and stylized silks.

There is a wide band supporting the headdress with an S-shaped element believed to represent water (Leigh 1970, 257) on each side of the "C" glyph. The face has the characteristic elements of the God of Glyph "L," including the aspas and buccal and nasal masks. The mouth is open, displaying teeth as commonly depicted on figurines from Oaxaca and a characteristic facial expression seen on many rural Zapotec speakers today (Brockington personal communication). This feature is evident on most urns. The figure wears the characteristic pectoral ornament, and although I am not aware that the glyph adorning the breast has been identified or explained, it is frequently seen on Zapotec urns. The figure wears elaborate earspools, a bracelet on each wrist, a feather cape around the shoulders, and a decorative loincloth. It is in the customary posture, seated cross-legged with the hands on the knees palms downward. This urn was illustrated in Caso and Bernal (1952, 93, fig. 153) before repairs were made to the piece. In figure 82 it appears in the restored state.

Figure 83 shows a closeup of the ear of maize on the left side of the headdress. Only a segment of a complete ear is represented; this is the way naturalistic representations are often depicted on Zapotec urns. When partial ears are exhibited, the identification of the race of maize is based on kernel size and shape, row number, and ear diameter. The ear on the right was restored and does not represent a naturalistic ear of maize.

Z-2. Fig. 84. Urn of the God of Glyph "L."
Museo Nacional de Antropología 6–2293.
Provenience: Tomb 40, Monte Albán.
Fabric: buff with a rough patina.
Dimensions: ht. 32.0 cm, width 30.0 cm.
Period: Monte Albán IIIB.
Race of maize: Nal Tel/Pira.

The God of Glyph "L" with all its characteristic attributes is shown in figure 84. The "C" glyph appears as the central medallion of the headdress and is surmounted by a penache flanked on either side by an ear of maize with stylized silks and staminate tips. The headband supporting the headdress has an S-shaped element on either side of the central medallion. The face has a nasal and buccal mask, and the mouth is open displaying teeth. The

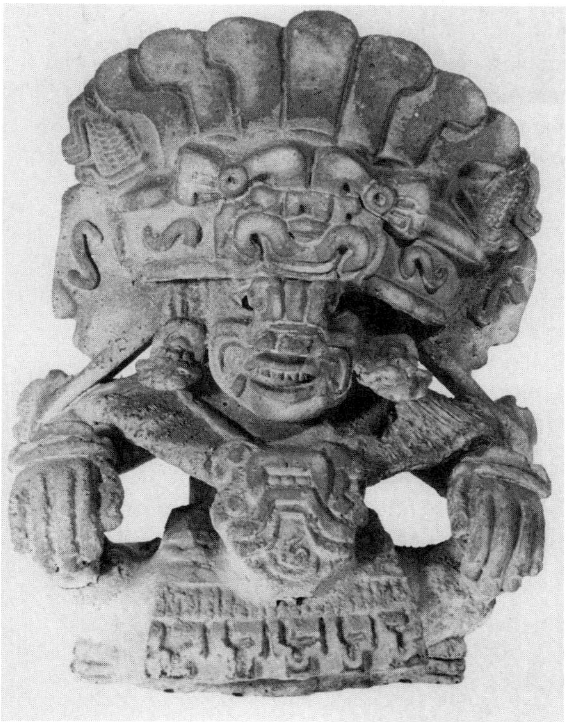

Fig. 84. Z-2. Urn of the God of Glyph "L."

figure wears a pectoral ornament, elaborate earspools, a bracelet on each wrist, a feather cape around the shoulders, and a loincloth. All elements are like the urn in figure 82, with the exception of the border elements on the loincloth and the race of maize appearing in the headdress. The border element on the loincloth is an inverted pyramid design that represents the Sky God (Howard Leigh personal communication). The race of maize in the headdress has been identified as Nal Tel. This description applies equally to Z-3 and Z-4, and they are three of five identical urns found in Tomb 40 at Monte Albán, illustrated by Caso and Bernal (1952, 92; figs. 152 and 153). The ears of maize in the headdresses of these urns were all impressed from the same mold.

Z-3. Urn of the God of Glyph "L."
Museo Nacional de Antropología 6–2290.
Provenience: Tomb 40, Monte Albán.
Fabric: buff with a rough patina.
Dimensions: ht. 32.0 cm, width 30.0 cm.
Period: Monte Albán IIIB.
Race of maize: Nal Tel/Pira.

Z-4. Urn of the God of Glyph "L."
Museo Nacional de Antropología 6–2289.
Provenience: Tomb 40, Monte Albán.
Fabric: buff with a rough patina.
Dimensions: ht. 32.0 cm, width 31.0 cm.
Period: Monte Albán IIIB.
Race of maize: Nal Tel/Pira.

Z-5. Fig. 85. Urn of the God of Glyph "L."
Museo Nacional de Antropología 6–6717.
Provenience: Patio 1 of Tomb 138, Monte Albán.
Fabric: buff with a rough patina.
Dimensions: ht. 28.3 cm, width 22.2 cm.
Period: Monte Albán IIIB.
Race of maize: Nal Tel/Pira.

The "C" glyph central medallion of the headdress is surmounted by a penache flanked on each side by an ear of maize with stylized silks and

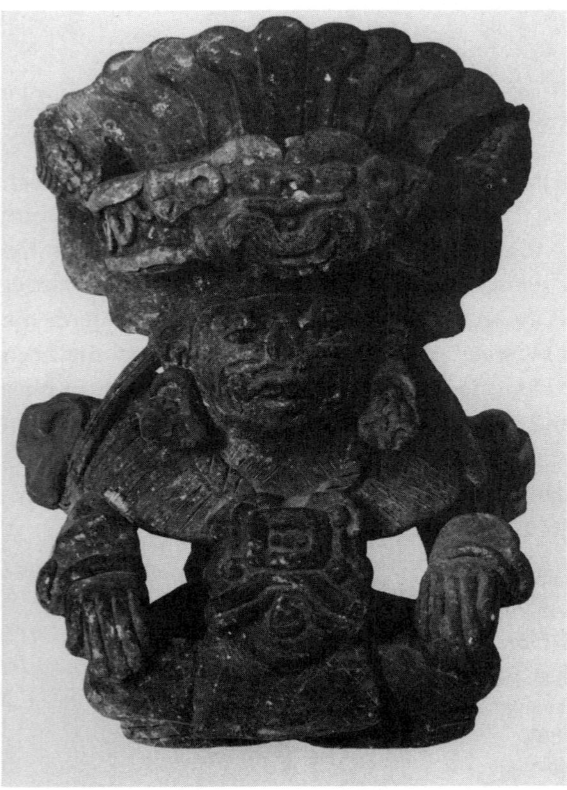

Fig. 85. Z-5. Urn of the God of Glyph "L."

staminate tip. The nose has been damaged, but the buccal mask and other elements associated with the deity are present. The figure wears the pectoral ornament, a bracelet on each wrist, and elaborate earspools seen on the four previous urns. He is seated in the characteristic cross-legged position and wears a feather cape and plain loincloth. This urn is also pictured in Caso and Bernal (1952, 94, fig. 155).

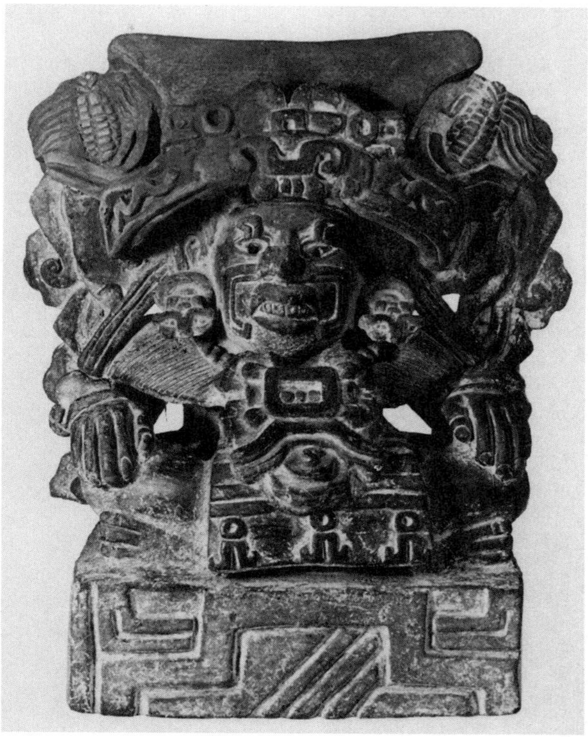

Fig. 86. Z-6. Urn depicting the God of Glyph "L."

Z-6. Fig. 86. Urn depicting the God of Glyph "L."
Museo Nacional de Antropología 6–6754.
Provenience: Antechamber of Tomb 31, Monte Albán.
Fabric: gray with buff.
Dimensions: ht. 30.0 cm, width 26.7 cm.
Period: Monte Albán IIIB.
Race of maize: Nal Tel/Pira.

This is one of five identical urns illustrated in situ by Caso and Bernal (1952, 93; fig. 154). This piece has been repaired, and the ear of maize on the left is not authentic. The figure has attributes associated with the God of Glyph "L" and a tassel element bordering the loincloth. This figure is unusual, however, because it is seated on a square pedestal with elaborate ornamentation on side panels in back of the figure, a throne-like structure referred to as a dossel. Two other examples of figures with dossels are illustrated by Boos (1966b, 185, 269, figs. 164 and 247).

Z-7. Fig. 87. Urn depicting the God of Glyph "L."
Museo Frissell de Arte Zapoteca 6883.
Provenience: Excavated at Lambityeco, Oaxaca.
Fabric: gray, friable.
Dimensions: ht. 18.0 cm, width 10.6 cm.
Period: Monte Albán IV.
Race of maize: Nal Tel/Pira.

The decline in the technical quality of ceramics that is diagnostic of Monte Albán IV is well illustrated by this specimen. There is a central medallion glyph in the headdress, but it cannot be identified. The headdress is com-

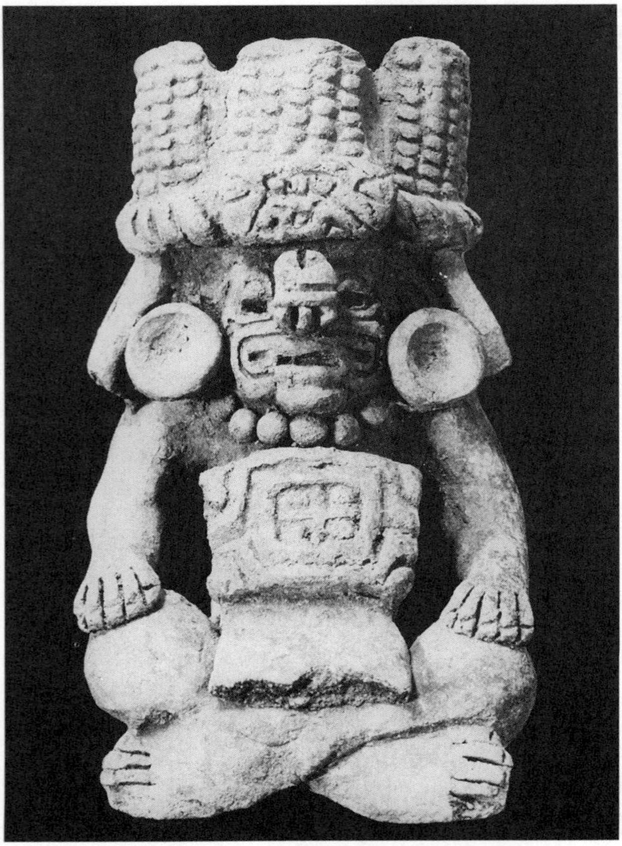

Fig. 87. Z-7. Urn depicting the God of Glyph "L."

posed of three standing ears of maize, but they are not full ears as might be expected. Although the center ear is wider as a result of having been flattened out when it was applied to the urn, it is clear that each ear came from the same mold. The face has nasal and buccal masks. The eyes, however, appear more characteristic of Cocijo than of the God of Glyph "L." The pectoral ornament displays the "E" glyph (Caso 1928, 32, fig. 8).

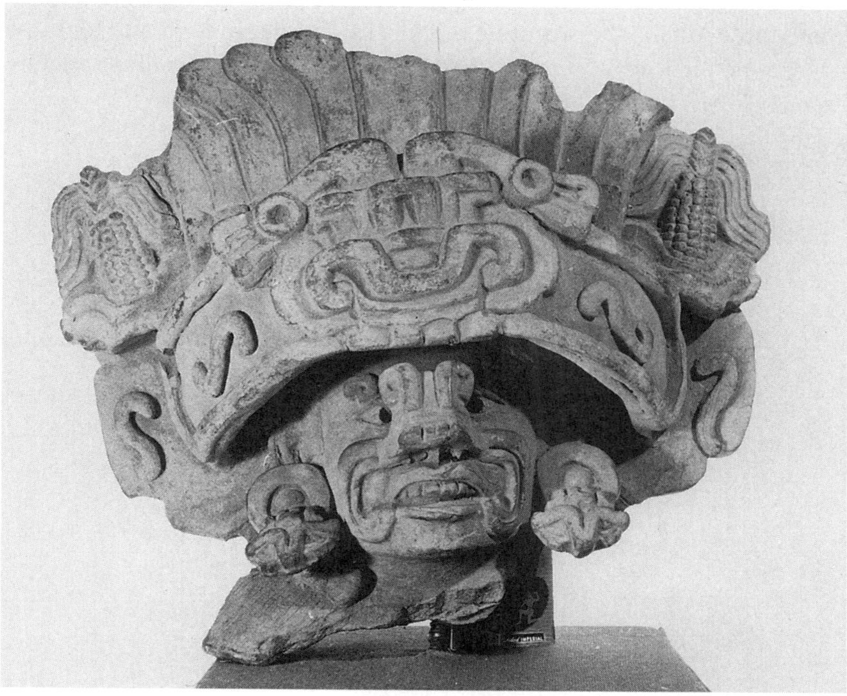

Fig. 88. Z-8. Head from an urn depicting the God of Glyph "L."

Z-8. Fig. 88. Head from an urn depicting the God of Glyph "L."
Museo Nacional de Antropología 6–338.
Provenience: Zaachila.
Fabric: buff.
Dimensions: ht. 30.0 cm, width 35.0 cm.
Period: Monte Albán IV.
Race of maize: Nal Tel/Pira.

The provenience of the head of the God of Glyph "L" in figure 88 is Zaachila, and ordinarily the antiquity of an urn from this provenience without documentation from excavation would be suspect. In this case, however, the attributes are all characteristic of the deity, and the piece is well executed in what is considered true Zapotec style. The "C" glyph is the central medallion of the headdress and is surmounted by a penache flanked on either side by an ear of maize with stylized silks and staminate tip. The headband has an S-element on each side of the central medallion. The face has the diagnostic nasal and buccal masks and aspas over the eyes, and the figure wears elaborate earspools like those appearing on other urns from Monte Albán. The execution is typically Zapotec in style, and for these reasons the piece must be considered authentic. This urn is illustrated by Caso (1928, 40, fig. 17).

Z-9. Fig. 89. Urn depicting the God of Glyph "L."
Smithsonian Institution Museum of Natural History 115032.
Provenience: Acatlán.
Fabric: buff.
Dimensions: ht. 18.8 cm, width 13.5 cm.
Period: Monte Albán IV.
Race of maize: unidentified.

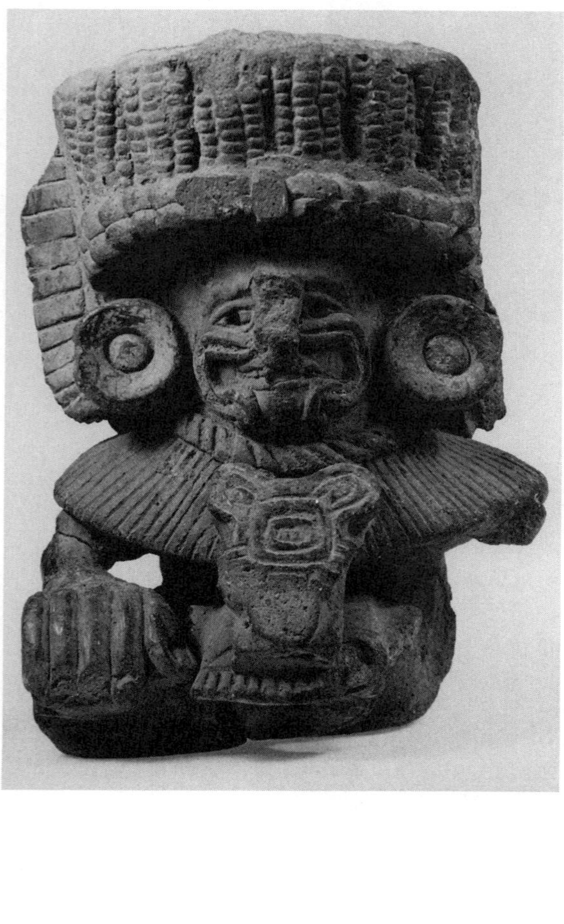

Fig. 89. Z-9. Urn depicting the God of Glyph "L."

Some of the characteristic attributes of the God of Glyph "L" seen in other urns are missing from the one in figure 89. This urn is part of a collection made by Louis Ayme, an American civil engineer who directed the construction of the railroad through Oaxaca in the mid-1800s. The urns in this collection are considered authentic because they are supposed to have been discovered during the construction of the railroad. The "C" glyph is notably absent from this urn, but this is not unusual in Monte Albán IV because the "C" glyph was rapidly falling into disuse. The headdress is formed by six vertical ears of maize. The nose is too badly damaged to tell whether the nasal mask was present, but the buccal mask is evident. It seems the aspas was omitted. The figure wears a pectoral ornament, large plain earspools, a feather cape, and a loincloth. It is seated cross-legged, and though the limbs are damaged, it can be seen that the hands were disproportionately large.

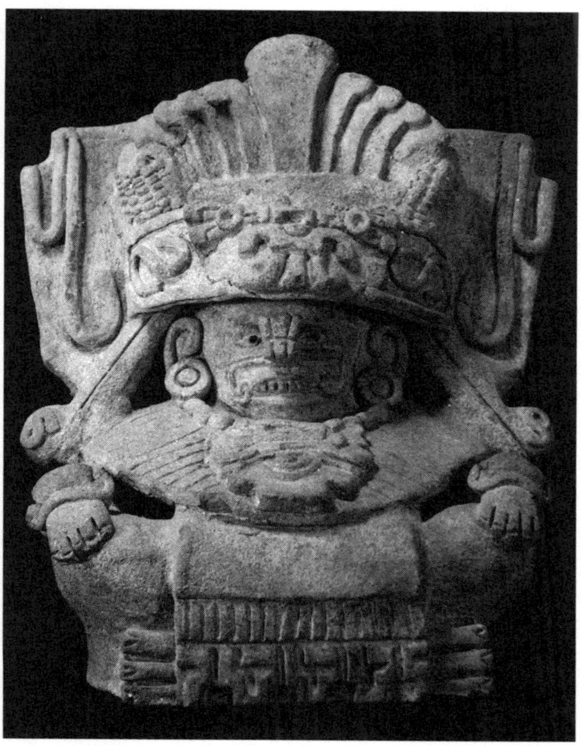

Fig. 90. Z-10. Urn depicting the God of Glyph "L."

Z-10. Fig. 90. Urn depicting the God of Glyph "L."
Museo Frissell de Arte Zapoteca 8870.
Provenience: unknown.
Fabric: gray with buff patina and faint traces of red.

Dimensions: ht. 26.3 cm, width 23.1 cm.
Period: Monte Albán IIIB/IV.
Race of maize: Nal Tel/Pira.

The characteristic "C" glyph in the headdress appears in a slightly modified form. It is surmounted by a penache flanked on either side by a single ear of maize that does not have the usual stylized silks and staminate tip. The headband has an S-element, but here it appears in exaggerated form. Side panels flanking the headdress also depict the S-element in a modified and greatly exaggerated form. This oddity, not seen on any other urn, should be carefully noted. The face exhibits the nasal and buccal masks, but the *aspas* are missing. Also, the treatment of the ears is unusual and unique as far as I know. The figure wears the pectoral glyph, small, round earspools, a bracelet on each wrist, a feather cape, and a loincloth bordered by inverted pyramids. I have assigned this urn to Monte Albán III-B or IV. The combination of attributes is appropriate for Monte Albán III-B, but they are later in style than most of the urns of that period and the piece is considered transitional to Monte Albán IV. This urn was illustrated by Paddock (1970, 215, fig. 268) and Boos (1966b, Lam. LV).

Z-11. Urn depicting the God of Glyph "L."
Museo Frissell de Arte Zapoteca S-1796.
Provenience: unknown.
Fabric: gray with buff patina and faint traces of red.
Dimensions: ht. 26.3 cm, width 22.8 cm.
Period: Monte Albán IIIB/IV.
Race of maize: Nal Tel/Pira.

This piece is identical to Z-10, and the same description applies.

Z-12. Fig. 91. Urn depicting the God of Glyph "L."
Museo Nacional de Antropología 6–5421.
Provenience: Miahuatlán.
Fabric: buff.
Dimensions: ht. 36.3 cm, width 33.1 cm.
Period: Monte Albán IIIB.
Race of maize: Nal Tel/Pira.

Since the provenience for this piece is Miahuatlán, a place where fakes were manufactured in large numbers, it has to be suspect. Most of the attributes are in order, however, and it may well be an authentic specimen. The "C" glyph forms the central medallion of the headdress and is surmounted by a penache flanked on either side by an ear of maize that can be identified as Nal Tel. The treatment of the maize, though, is unusual because there are

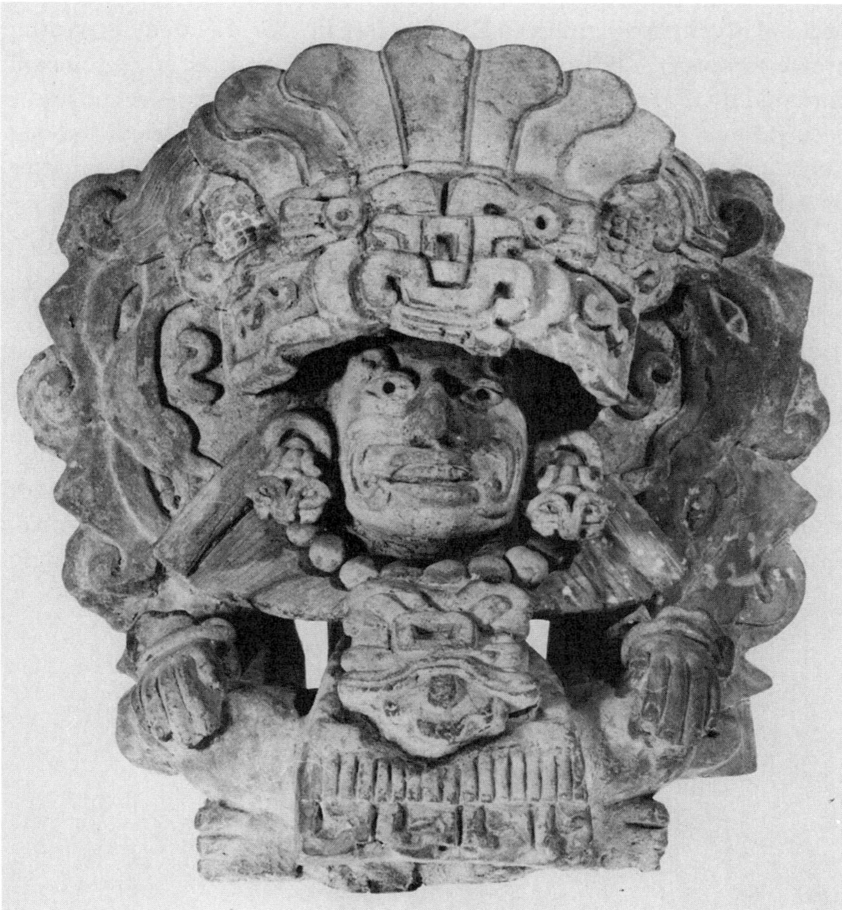

Fig. 91. Z-12. Urn depicting the God of Glyph "L."

stylized silks but no staminate tips. Also, there is a semivolute element curling outward at the base of the ear and at the tip. I have not seen maize depicted like this on any other specimen. It would be impossible to say the piece is fake on the basis of this evidence alone because it could be merely a departure from convention by the individual artist. The headband on this piece has the symbol for undulating water, but in this case, it is not the characteristic S-shape element seen previously. Here the wave has two full loops. The face has the aspas and buccal mask. The nose has been damaged and repaired; the nasal mask may have originally been present. The figure wears a pectoral glyph, but it is not the usual form. It closely resembles the pectoral glyph on an urn from Zaachila (see fig. 88). The deity is wearing ornate earspools, a bracelet on each wrist, a feather cape, and a loincloth bordered by the inverted pyramid design. It is seated cross-legged and is backed by an elaborately carved dossel. A similar dossel is depicted on a Cocijo urn at the Museum für Völkerkunde Berlin that TL dating determined is post-Columbian (see Schuler-Schömig 1970).

Z-13. Fig. 92. Urn depicting the God of Glyph "L."
Museo Nacional de Antropología 6–6757.
Provenience: Zaachila.
Fabric: gray with white traces.
Dimensions: ht. 8.9 cm, width 8.1 cm.
Period: Monte Albán IV.
Race of maize: Nal Tel/Pira.

This unusual jar had been assigned to the Monte Albán II–III transition according to the museum catalog. This would be centuries before molds were used for manufacturing urns. The face has the buccal mask but lacks

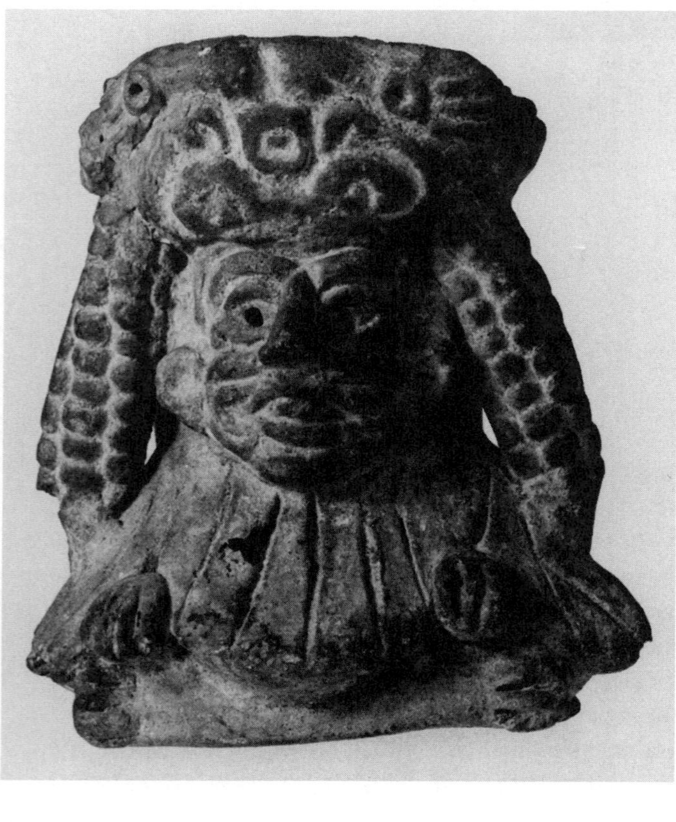

Fig. 92. Z-13. Urn depicting the God of Glyph "L."

the nasal mask and aspas, and the figure wears a feather cape. Overall, it is poorly executed and on this basis would be assigned to Monte Albán IV. The maize on this piece is also unusual. A full ear flanks each side of the figure, seemingly with the butt end emanating from the headdress and pointing downward. It is an eight-rowed ear, and the isodiametric kernels are in single alternating rows.

Z-14. Fig. 93. Urn depicting the God of Glyph "L."
Yale University Art Gallery 1973.88.10.
Provenience: unknown.
Fabric: gray.
Dimensions: ht. 43.2 cm, width 43.8 cm.
Period: Monte Albán IIIB.
Race of maize: Nal Tel/Pira.

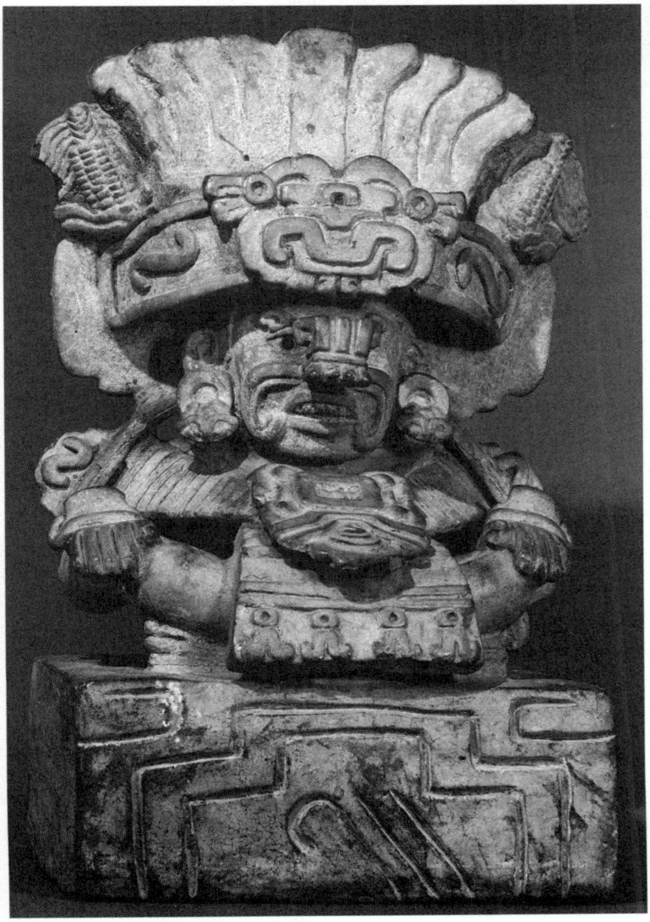

Fig. 93. Z-14. Urn depicting the God of Glyph "L." Yale University Art Gallery Stephen Carlton Clark, B.A. 1903, Fund.

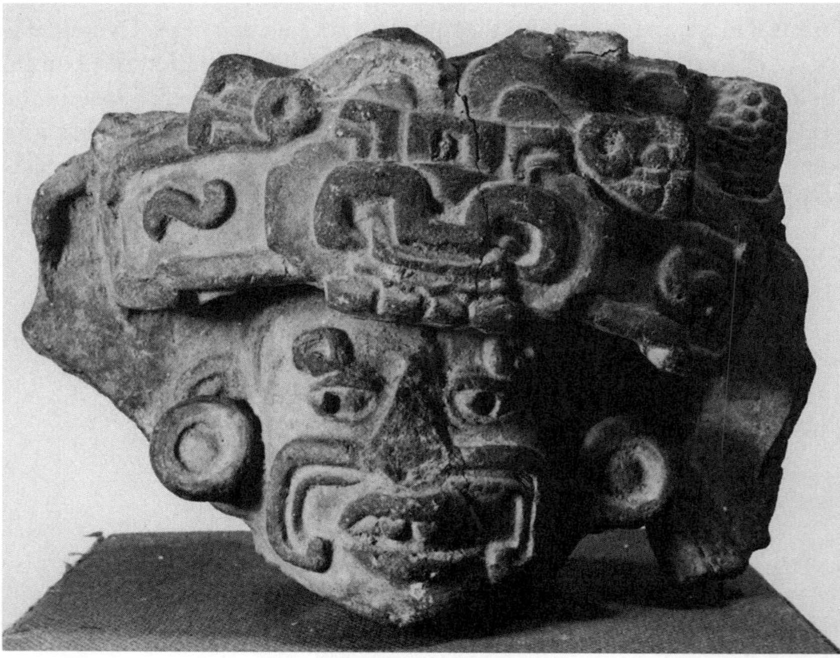

Fig. 94. Z-15. Head fragment depicting the God of Glyph "L."

The figure is seated cross-legged with hands on knees on a throne-like platform that bears the Zapotec glyph for high places. It wears the characteristic nasal and buccal masks of the God of Glyph "L," and the central medallion of the headdress is the "C" glyph. The figure wears elaborate earspools, a glyph pectoral ornament, wrist bands, a feather cape, and a loincloth with a border of four tassel-like elements. The "C" glyph in the headdress is surmounted by a penache flanked on either side by an ear of maize with stylized silks and staminate tip.

Z-15. Fig. 94. Urn of the God of Glyph "L."
Museo Nacional de Antropología 6–6736.
Provenience: unknown.
Fabric: buff.
Dimensions: ht. 15.0 cm, width 21.2 cm.
Period: Monte Albán IIIB.
Race of maize: Nal Tel/Pira.

This piece has been badly damaged, and all that remains is the head, which has been repaired. The nose has been broken off. The eyes have holes in place of the pupils and are surmounted by eyebrows. The face has the characteristic "L" elements on the cheeks, and the mouth is open, showing teeth. The headdress band is adorned with the central "C" glyph, which is flanked by S-elements symbolizing water on each side. There was a central penache in the headdress flanked by an ear of maize on each side. The maize ear on the figure's left is intact and identifiable. The figure wears plain round earspools.

Z-16. Fig. 95. Urn depicting the God of Glyph "L."
University of Pennsylvania University Museum NA2141.
Provenience: unknown.
Fabric: gray.
Dimensions: ht. 28.1 cm, width 18.1 cm.
Period: Monte Albán IIIB.
Race of maize: Chapalote.

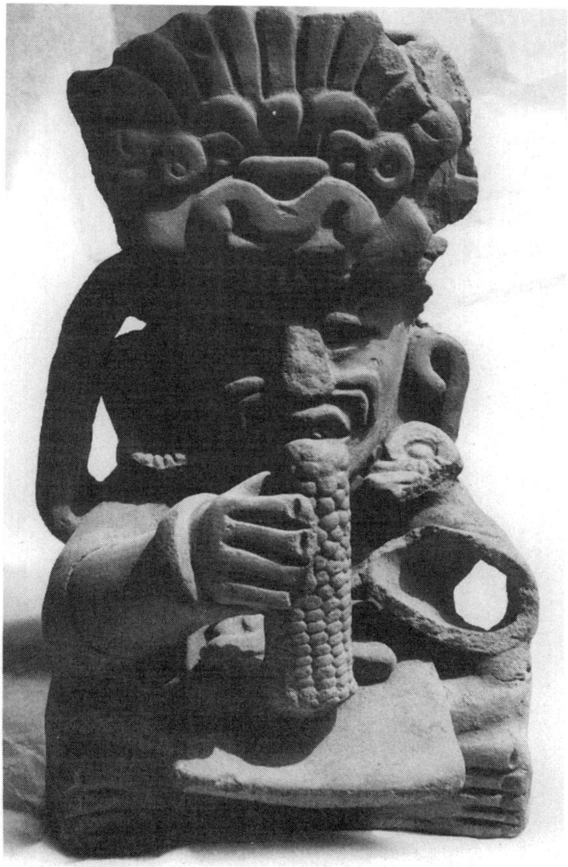

Fig. 95. Z-16. Urn depicting the God of Glyph "L."

The figure is seated cross-legged in typical fashion, but rather than having its hands on its knees, it holds an ear of maize in front. The urn has been damaged and one arm is missing. The characteristic buccal mask is portrayed, but the nose is broken so it may or may not have displayed the nasal mask. The "C" glyph is the central medallion of the headdress, but both sides of the headdress are broken off. The figure wears a plain loincloth, elaborate earspools, a pectoral ornament, and a bracelet.

Z-17. Fig. 96. Urn depicting the God of Glyph "L."
Museo Nacional de Antropología 6–275.
Provenience: Zaachila.
Fabric: dark gray.
Dimensions: ht. 26.7 cm, width 19.4 cm.
Period: Monte Albán IV.
Race of maize: Nal Tel/Pira.

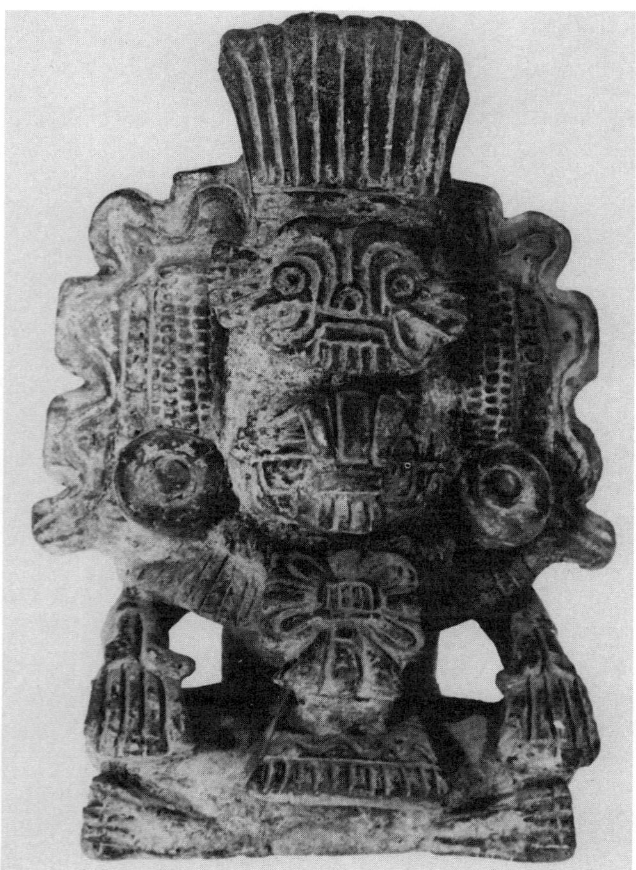

Fig. 96. Z-17. Urn depicting the God of Glyph "L."

In addition to a provenience of Zaachila, a town that housed a factory that made fakes, there are stylistic reasons why this urn is suspect. The "C" glyph central medallion of the headdress is surmounted by a penache that does not extend over the entire headdress as is usually seen, and the plumes are narrow vertical straight lines, not nicely shaped as is the custom. The face has a large nasal mask, and the buccal mask is present. The glyph on the pectoral ornament cannot be identified and is unlike any depicted in the literature. The maize is out of character because a single ear appears extended vertically on each side of the face, a scalloped decoration surrounds the back of the maize, and the figure's face extends from the base of the penache to the shoulders. The execution of the entire piece is atypical of Zapotec art style and is poorly made. The urn is illustrated by Caso and Bernal (1952, 94, plate 156).

Z-18. Urn depicting the God of Glyph "L."
Museo Nacional de Antropología 6–328.
Provenience: Zaachila.
Fabric: dark gray.
Dimensions: ht. 24.0 cm, width 19.5 cm.
Period: Monte Albán IV.
Race of maize: Nal Tel/Pira.

This piece has been broken and the lower portion is missing. All features present, however, are just like those of Z-17 and the same description applies.

Z-19. Fig. 97. Urn depicting the God of Glyph "L."
St. Louis Art Museum 68:452.
Provenience: unknown.
Fabric: dark gray, traces of red paint.
Dimensions: ht. 17.3 cm, width 14.5 cm.
Period: Monte Albán IV.
Race of maize: Nal Tel/Pira.

This urn resembles Z-17 and Z-18 in overall style. There are differences in the "C" glyph in the headdress and the pectoral glyph.

Z-20. Urn depicting the God of Glyph "L."
St. Louis Art Museum 3051.73.
Provenience: unknown.
Fabric: dark gray, traces of red paint.
Dimensions: ht. 33.9 cm.
Period: Monte Albán IV.
Race of maize: Nal Tel/Pira.

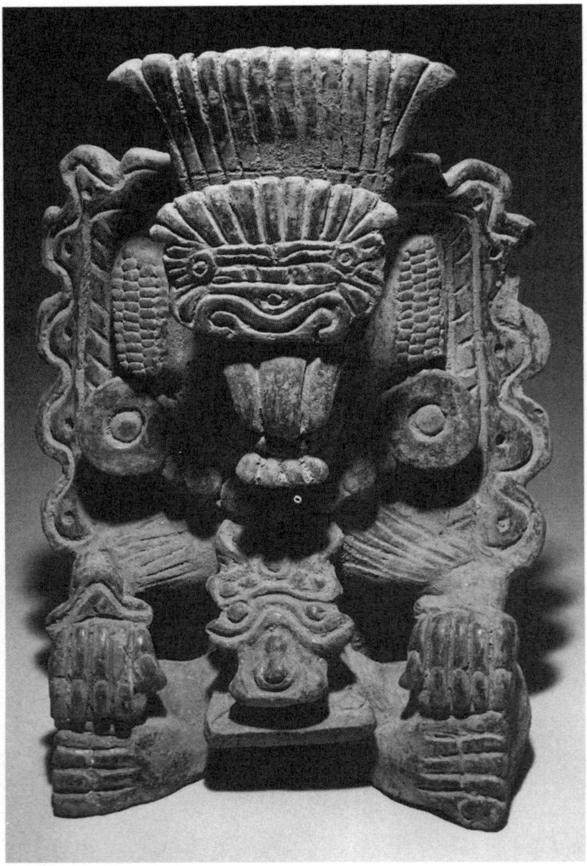

Fig. 97. Z-19. Urn depicting the God of Glyph "L."

This urn resembles Z-17, Z-18, and Z-19 in overall style. But the figure wears a cord necklace without a pectoral glyph, it holds a vessel with two stylized leaf-like elements emerging from it with both hands, and it has a plumed penache.

Z-21. Urn depicting the God of Glyph "L."
Museum für Völkerkunde Munich 10.3505.
Provenience: unknown.
Fabric: gray with buff patina.
Dimensions: ht. 22.2 cm, width 15.8 cm.
Period: Monte Albán IV.
Race of maize: Nal Tel/Pira.

The nasal and buccal masks are missing, but otherwise the piece closely resembles Z-18 and Z-19.

Z-22. Urn depicting the God of Glyph "L."
Museum für Völkerkunde Munich 14.8.6.
Provenience: Tuquila.
Fabric: gray with buff patina and traces of red paint.
Dimensions: ht. 30.6 cm, width 23.1 cm.
Period: Monte Albán IV.
Race of maize: Nal Tel/Pira.

The bottom portion of this figure is broken, but the upper body is intact. The figure has some unusual attributes not seen on other urns. There is a "C" glyph as central medallion of the headdress with six partial ears of maize behind it. At the base of the maize ears are two rows of round elements with incised circles inside each element. The six maize ears are surmounted by a penache that has a rope-like element edging it. The face has nasal and buccal masks with a highly unusual element of incised and punctate lines that emerge from the mouth. The figure wears round earspools that have atypical tassels suspended from them and a necklace of alternating round and oblong beads. The combination of atypical attributes makes the antiquity of this piece suspect.

Z-23. Urn depicting the God of Glyph "L."
Museum für Völkerkunde Munich 10.3506.
Provenience: unknown.
Fabric: gray with buff patina.
Dimensions: ht. 18.3 cm, width 12.2 cm.
Period: Monte Albán IV.
Race of maize: Nal Tel/Pira.

This piece resembles Z-22 in style and has other atypical attributes. It has the "C" glyph as the central medallion of the headdress that is unlike other "C" glyphs I have seen because an array of feathers in a sunburst pattern emanates directly from it. The figure wears a rope necklace and tassel ornaments on each arm. In addition to the four standing ears of maize in the headdress, it has a mold-made ear of maize as a pectoral ornament.

Z-24. Fig. 98. Urn depicting the God of Glyph "L."
American Museum of Natural History 30.3/2301A.
Provenience: unknown.
Fabric: gray.
Period: Monte Albán IIIB/IV.
Dimensions: ht. 40.0 cm, width 30.0 cm.
Race of maize: Mochero.

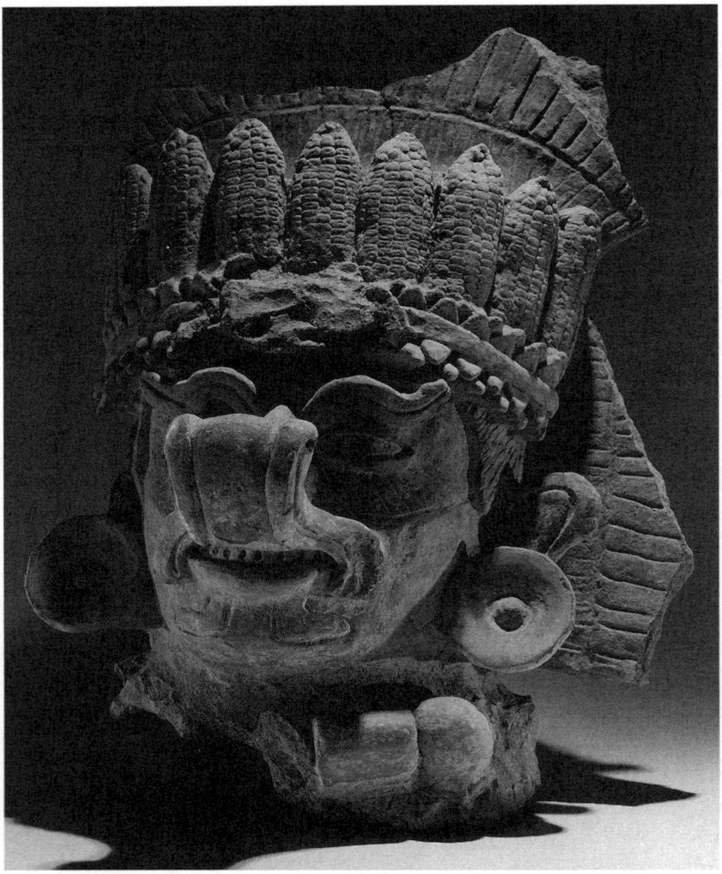

Fig. 98. Z-24. Head fragment depicting the God of Glyph "L."

This is a head of the God of Glyph "L" that was part of a large urn. Although it has been damaged, the face and nine standing ears of maize in the headdress are intact. The central glyph of the headdress is missing, and most of the penache surmounting the maize ears is broken off. The figure has nasal and buccal masks and eye elements characteristic of the God of Glyph "L." It wears round earspools, and remnants of a bead necklace are evident. The maize depicted is a type of dent corn that some scientists consider modern. This piece was dated by thermoluminescence, and results confirmed that it was fired in antiquity (David Zimmerman personal communication, August 4, 1975). Although the Zimmerman thermoluminescence tests have been criticized because thorium levels were estimated to be higher than they actually are in some of the clays (Elizabeth P. Benson personal communication, December 31, 1982), retesting was never done because of David Zimmerman's untimely death. Because results from subsequent TL testing of Zapotec urns in the Berlin collection by a European lab are in line with Zimmerman's conclusions, this specimen seems to confirm the presence of dent corn in Oaxaca by A.D. 900.

The race of maize that is the closest match for this piece is the Peruvian race Mochero. Since modern Mochero does not have much denting of the kernels that is distinctly displayed on this piece, however, the alternative choices include Aleman of Peru and Tuxpeño of Mexico. Although Aleman is believed to have been a nineteenth-century introduction into Peru by German settlers, it is derived from the Mexican race Tuxpeño (Grobman et al. 1961, 309). Tuxpeño is believed to have originated from hybridization between the Mexican races Tepecintle and Olotillo (Wellhausen et al. 1952, 149–54). This theory supports the contention that these races date to prehistoric times and indicates that Tuxpeño was in existence by at least A.D. 500. This is significant because Tuxpeño is ancestral to the most productive agricultural races, including the southern dent corns of the United States.

Z-25. Fig. 99. Urn depicting the God of Glyph "L."
American Museum of Natural History 30.2/6099.
Provenience: unknown.
Fabric: gray with traces of blue and red.
Period: Monte Albán IIIB.
Dimensions: ht. 29.4 cm, width 20.0 cm.
Race of maize: Nal Tel/Pira.

The figure of the God of Glyph "L" is seated cross-legged with palms on knees, wearing a loincloth and a feather cape. He wears round earspools and a cord necklace with pectoral glyph. The face has the characteristic nasal and buccal masks. There are six standing ears of maize in the headdress surmounted by a penache. The urn is washed with blue and red colors that are authentic Maya blue pigment and cinnabar, respectively.

Z-26. Fig. 100. Urn depicting the God of Glyph "L."
American Museum of Natural History 30.0/3.
Provenience: unknown.
Fabric: buff.
Dimensions: ht. 37.5 cm, width 25.0 cm.
Period: Monte Albán IIIB.
Race of maize: Karapampa.

The God of Glyph "L" figure is seated cross-legged with palms on the knees on a throne-like platform. It wears a feather cape and loincloth bordered by four tassel elements, bracelets on each wrist, elaborate ear ornaments, and a pectoral glyph. It has the characteristic nasal and buccal masks and eyebrow elements. A single ear of maize with stylized silks stands in the center of the headdress. The headband is ornamented with S-shaped water elements.

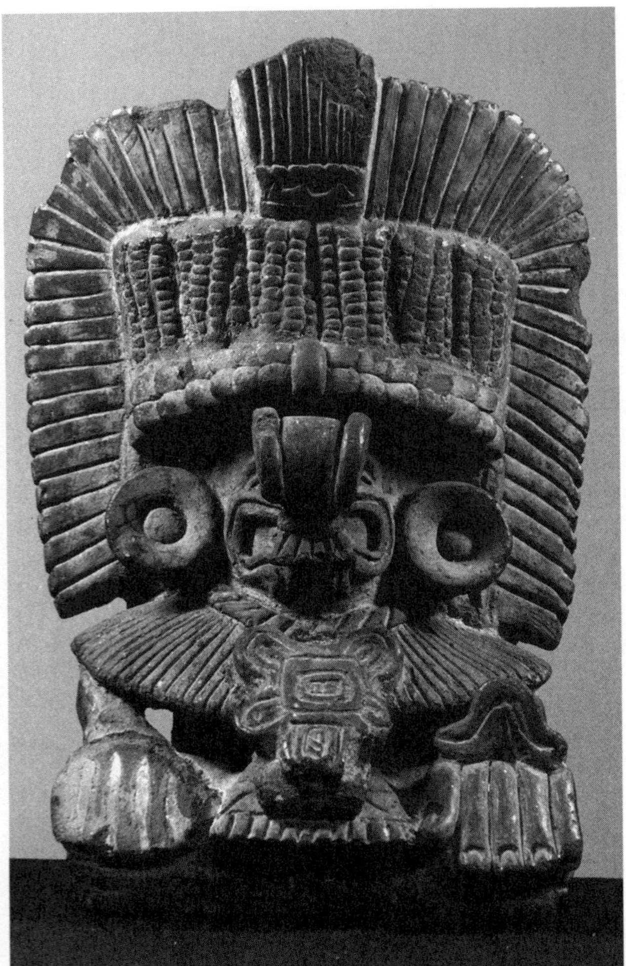

Fig. 99. Z-25. Urn depicting the God of Glyph "L."

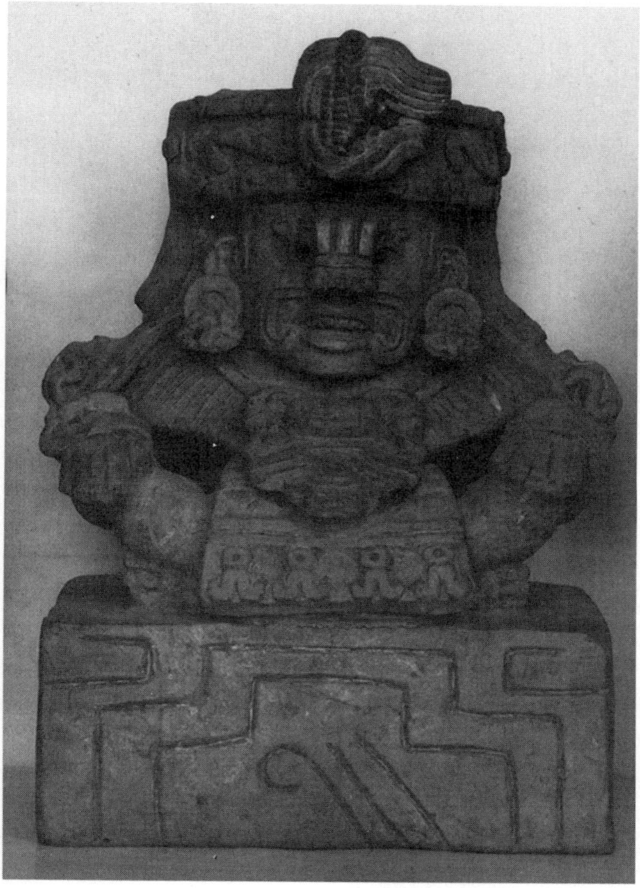

Fig. 100. Z-26. Urn depicting the God of Glyph "L."

Urns depicting the God Cocijo

Z-27. Fig. 101. Urn depicting the God Cocijo.
Smithsonian Institution Museum of Natural History 214981.
Provenience: unknown.
Fabric: black with light brown traces.
Dimensions: ht. 21.3 cm, width 13.8 cm.
Period: Monte Albán IV.
Race of maize: unidentified.

Since this piece does not display the characteristic "C" glyph, it deviates from the classic Cocijo form. Two vertical ears of maize with stylized silks and staminate tips are in the headdress. It appears that there were originally three such ears. The race of maize looks like Nal Tel/Pira but has not

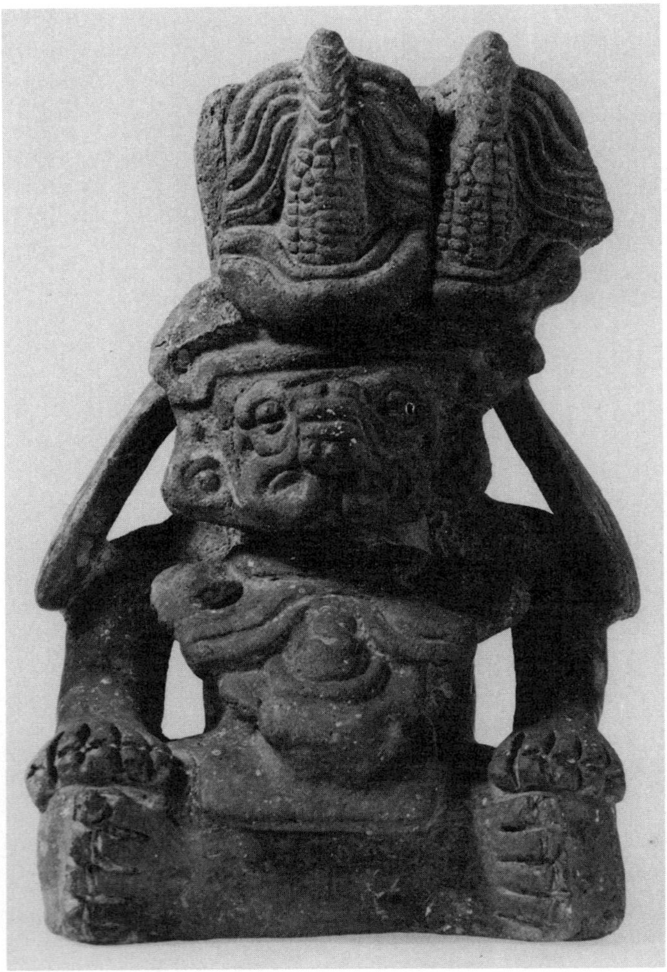

Fig. 101. Z-27. Urn depicting the God Cocijo.

been identified because necessary morphometric measurements are lacking. The figure displays the characteristic facial mask, wears a pectoral ornament, plain loincloth, small, plain, round earspools, and is seated cross-legged. It is illustrated by Boos (1966a, 23, fig. 46).

Z-28. Figs. 102 and 103. Urn depicting the God Cocijo.
Museo Nacional de Antropología 6–6708.
Provenience: The Mixteca.
Fabric: brown.
Dimensions: ht. 13.0 cm, width 11.3 cm.
Period: Monte Albán IV.
Race of maize: Nal Tel/Pira.

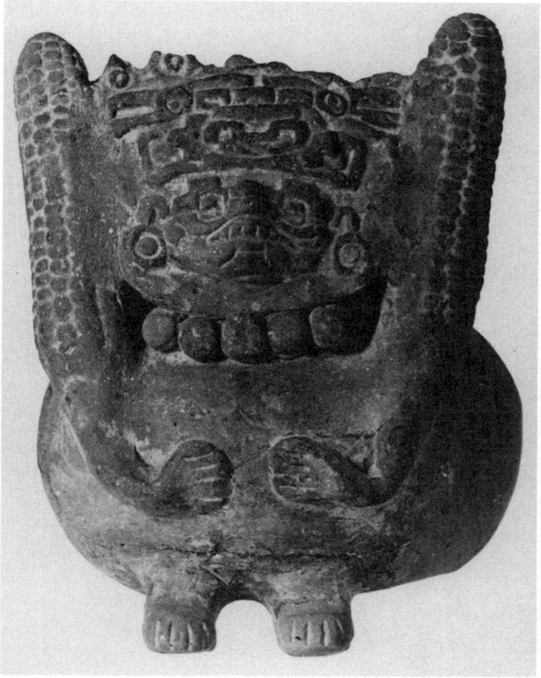

Fig. 102. Z-28. Urn depicting the God Cocijo.

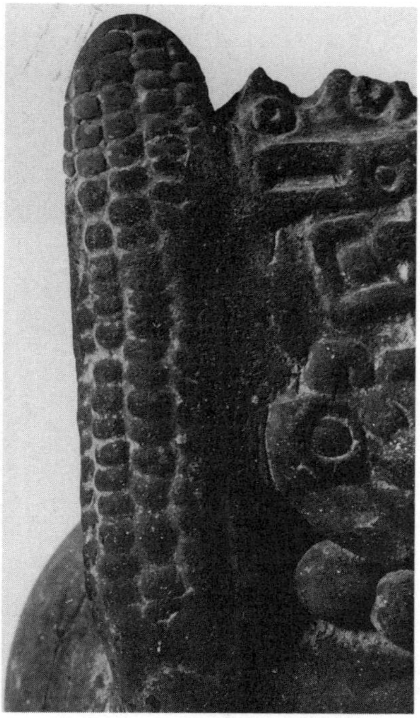

Fig. 103. Z-28. Closeup of maize ear on side of jar in fig. 102.

This urn deviates from the usual cylindrical shape and is more appropriately termed an olla, a globular form that was produced in Monte Albán IV (Caso and Bernal 1952, 40–42, figs. 46–50). The "C" glyph is of a later form, and the face does not have the characteristic bifurcated tongue. The face and headdress are a single molded piece that was attached to the jar. This technique was widely employed during Monte Albán IV. A vertical ear of maize is attached to each side of the vessel.

Z-29. Fig. 104. Urn depicting the God Cocijo.
Museo Nacional de Antropología 6–6758.
Provenience: Nochixtlán.
Fabric: buff.
Dimensions: ht. 15.5 cm, width 11.0 cm.
Period: Monte Albán IV.
Race of maize: Nal Tel/Pira.

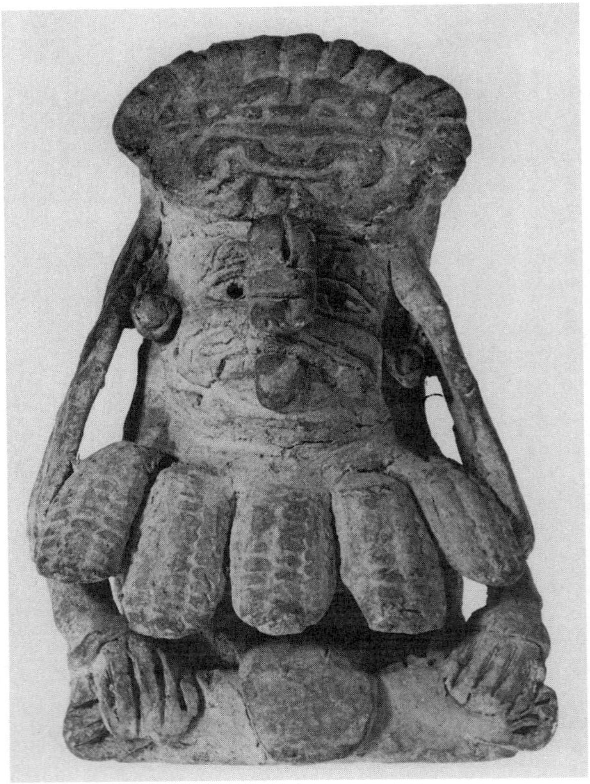

Fig. 104. Z-29. Urn depicting the God Cocijo.

This urn from the Mixteca region is chronologically late, as evidenced by the poor ceramic quality. The "C" glyph forms almost the entire headdress. The face has nasal and buccal masks. The extended tongue, however, is not bifurcated. The placement of the maize ears on this piece is unusual. There are five ears suspended from the neck in necklace fashion, appearing like a cape. This urn is illustrated in Caso and Bernal (1952, 100, fig. 106).

Z-30. Fig. 105. Urn depicting the God Cocijo.
Museo Nacional de Antropología 6–6013.
Provenience: unknown.
Fabric: brown with traces of white.
Period: Monte Albán IIIB/IV.
Race of maize: unidentified.

This urn has been classified as Cocijo because it bears all characteristic facial elements, but it is a composite of the God Cocijo and the Deity with the Ear of Maize in His Headdress. In the figure's headdress is displayed the

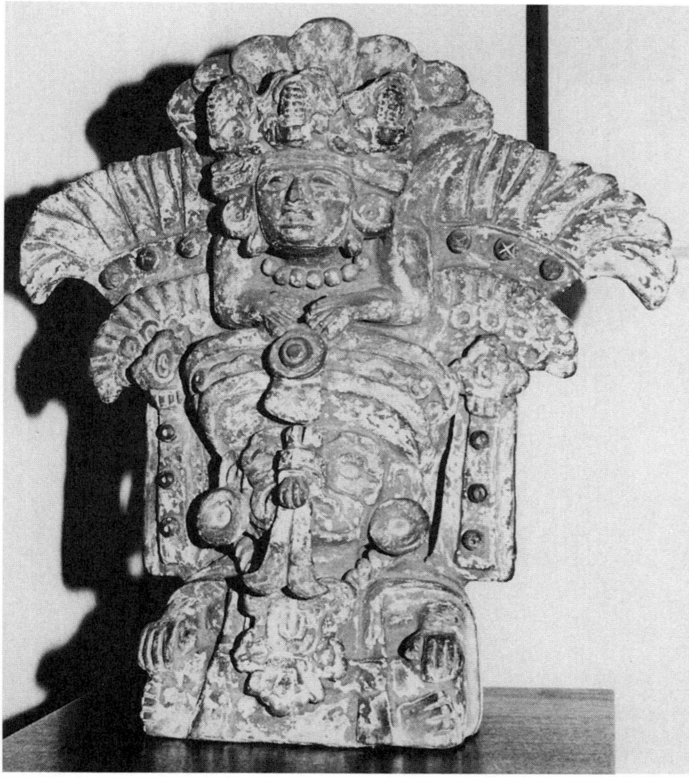

Fig. 105. Z-30. Urn depicting the God Cocijo.

headdress of Cocijo and three small maize ears with stylized silks but lacking staminate tips. It is a most unusual specimen. A drawing of a similar Cocijo urn is shown in Caso and Bernal (1952, 19, fig. 5). The difference is that the figure in the headdress of that urn is holding an ear of maize between his two hands and has a "C" glyph as the central medallion of the headdress. This piece was photographed on display in the Department of Archaeology at the Museo Nacional de Antropología in Mexico City. Urn dimensions, morphometric measurements, and catalog information were unavailable. Therefore, the race of maize has not been identified.

Z-31. Fig. 106. Urn depicting the God Cocijo.
Museo Nacional de Antropología 6–6672.
Provenience: Zaachila.
Fabric: gray.
Dimensions: ht. 37.4 cm, width 40.3 cm.
Period: Monte Albán IV.
Race of maize: Nal Tel/Pira.

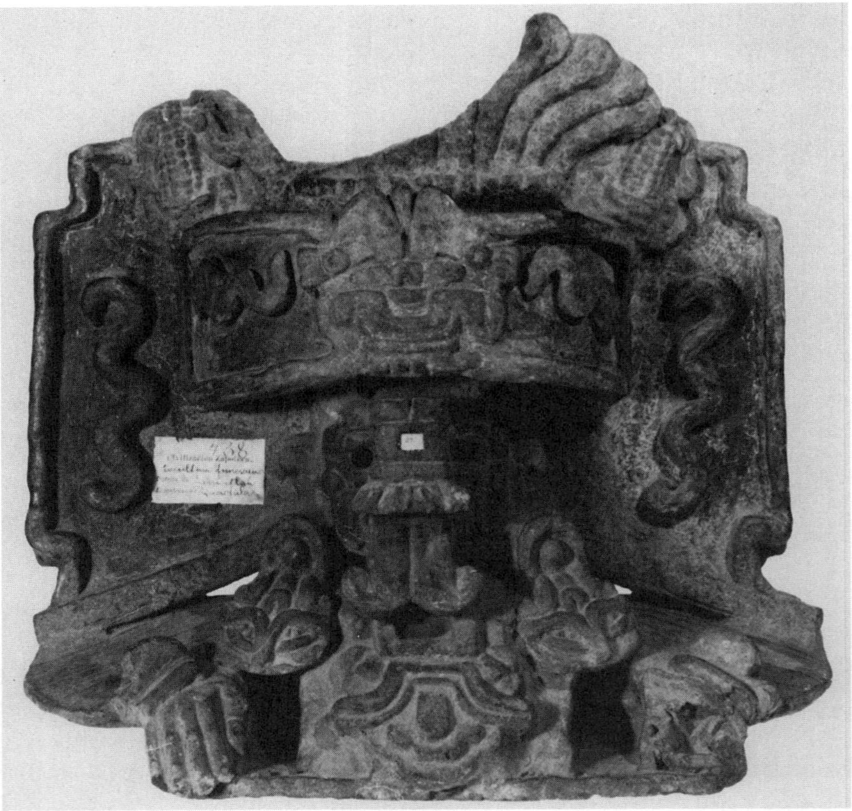

Fig. 106. Z-31. Urn depicting the God Cocijo.

This figure appears to have been a brazier cover because of the flat bottom and the lack of feet, which might have been depicted on the box. The back of the figure, however, forms the cylindrical body of an urn. It has a "C" glyph as the central medallion of the headdress surmounted by a penache and flanked on each side by an ear of maize with stylized silks and staminate tips. The water symbols occur on the headband and are composed of two full loops instead of the usual S-shape. The face has a large nasal mask, bifurcated tongue, and square eye elements. The figure wears a pectoral glyph, but the entire glyph is partially covered by the projecting tongue. The figure wears elaborate earspools that have been modeled by hand rather than molded, as is the usual treatment. The hands are depicted without arms, and parts of a feather cape can be noted. This piece is disproportionately wide for use as even a brazier cover. Urns that were used as brazier covers are unusual, but they do occur (Caso and Bernal 1952, 44, fig. 56).

Z-32. Urn depicting the God Cocijo.
Museo Nacional de Antropología 6–6674.
Provenience: unknown.
Fabric: gray.
Dimensions: ht. 38.7 cm, width 39.9 cm.
Period: Monte Albán IV.
Race of maize: Nal Tel/Pira.

This piece is virtually identical to Z-31, and the same description applies.

Z-33. Urn depicting the God Cocijo.
Museo Nacional de Antropología 6–6697.
Provenience: unknown.
Fabric: gray.
Dimensions: ht. 40.6 cm, width 38.1 cm.
Period: Monte Albán IV.
Race of maize: Nal Tel/Pira.

This piece is virtually identical to Z-31 and Z-32, and the same description applies.

Z-34. Fig. 107. Urn depicting the God Cocijo.
Museo Frissell de Arte Zapoteca 4279.
Provenience: Zaachila.
Fabric: dark reddish-brown.
Dimensions: ht. 10.6 cm, width 8.5 cm.
Period: Monte Albán IV.
Race of maize: Nal Tel/Pira.

This *florero* is thought to belong to Monte Albán IV, although the double-spout vase first appears in Monte Albán IIIA (Caso and Bernal 1965, 887). The spouts on each side of the vessel are ears of maize. The figure has the characteristic Cocijo mask, and the central medallion of the headdress is the face of the Jaguar God surmounted by a plume of upright trifoliate elements (Boos 1966b, 15). The jar itself forms the figure's body, with arms and feet. This is an unusual treatment for Zapotec pottery, but it is characteristic of Moche jars from the North Coast of Peru.

Z-35. Urn depicting the God Cocijo.
Museum für Völkerkunde Munich 04.95.
Provenience: unknown.
Fabric: dark gray.

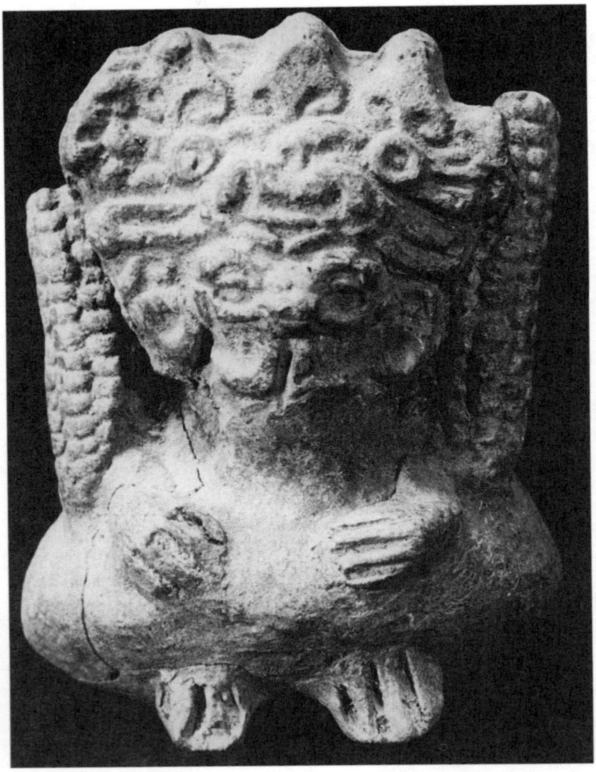

Fig. 107. Z-34. Urn depicting the God Cocijo.

Dimensions: ht. 21.0 cm, width 16.8 cm.
Period: Monte Albán IV.
Race of maize: Nal Tel/Pira.

This figure of Cocijo has two significant departures from convention. Here the projecting bifurcated tongue is depicted with the mouth open and lips pulled back showing the teeth. Furthermore, the hands are smaller than usually depicted by Zapotec artists, and they show all five fingers, which is rare on authentic pieces.

Z-36. Fig. 108. Brazier cover from urn depicting the God Cocijo.
American Museum of Natural History 30/6334.
Provenience: Xoxocotlán.
Fabric: buff with red.
Dimensions: ht. 47.5 cm, width 55.0 cm.
Period: Monte Albán IIIB.
Race of maize: Confite Puntiagudo/Canguil.

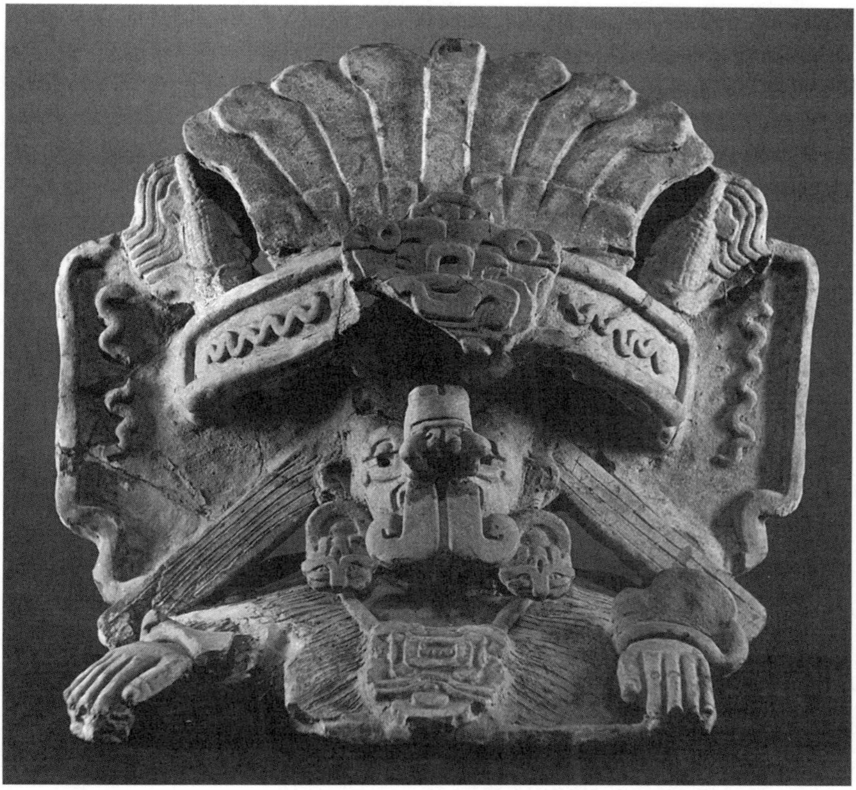

Fig. 108. Z-36. Brazier cover from urn depicting the God Cocijo.

This brazier cover is one in a series of five identical urns that were found cemented in the facade of Tomb 3 in Mound 9 at Xoxocotlán. The Cocijo figure wears a feather cape, ornate earspools, and nasal and buccal masks with the characteristic bifurcated tongue. Two standing ears of maize with staminate tips and stylized silks flank a "C" glyph central medallion in the headdress, all of which is surmounted by a penache. There is a dossel behind the figure.

Z-37. Urn depicting the God Cocijo.
American Museum of Natural History 30/6335.
Provenience: Xoxocotlán.
Fabric: buff with red.

Dimensions: ht. 47.5 cm, width 55.0 cm.
Period: Monte Albán IIIB.
Race of maize: unidentified.

This urn is one of the mates in the series of five urns found attached to the facade of Tomb 3 at Xoxocotlán. It is virtually identical to the urn described for Z-36 and illustrated in figure 108. The maize ears were too eroded to get accurate measurements, but since the ears are the same as those in Z-36, it is surmised that the race of maize is Confite Puntiagudo.

Z-38. Urn depicting the God Cocijo.
American Museum of Natural History 30/95.
Provenience: unknown.
Fabric: buff.
Dimensions: ht. 26.3 cm, width 17.5 cm.
Period: Monte Albán IIIB.
Race of maize: Nal Tel/Pira.

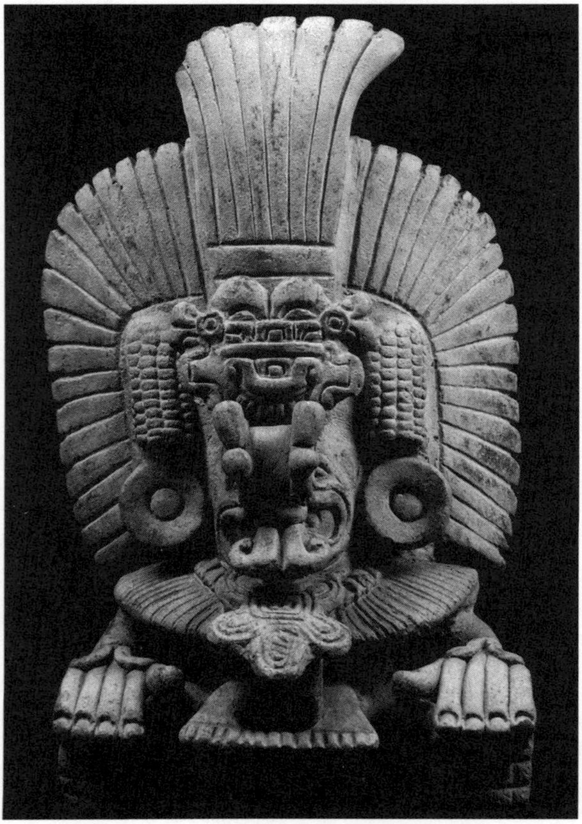

Fig. 109. Z-39. Urn depicting the God Cocijo from Zimatlán.

This Cocijo figure is seated cross-legged with palms on knees. It wears a loincloth and a feather cape, bead necklace, and round earspools, and it has the characteristic bifurcated tongue. Two ears of maize in the headdress flank a central "C" glyph and are surmounted by a penache.

Z-39. Fig. 109. Urn depicting the God Cocijo.
American Museum of Natural History 30/11173.
Provenience: Zimatlán.
Fabric: buff.
Dimensions: ht. 30.0 cm, width 18.8 cm.
Period: Monte Albán IIIB.
Race of maize: Nal Tel/Pira.

This Cocijo figure is seated cross-legged with palms on knees. It wears a loincloth and a feather cape, round earspools, and pectoral glyph and has the characteristic nasal and buccal masks with bifurcated tongue. There are two ears of maize in the headdress flanking a central "C" glyph surmounted by a penache.

Urn depicting the God Displaying the Mask of the God Cocijo in His Headdress

Z-40. Fig. 110. Urn depicting the God displaying the mask of the God Cocijo in his headdress.
Museo Nacional de Antropología Lola Olmeda de Olvera Collection.
Provenience: unknown.
Dimensions: ht. 52.0 cm.
Period: Monte Albán IIIB.
Race of maize: unidentified.

This elaborate piece is representative of the Zapotec culture at Monte Albán at its peak and is ceramic sculpture at its finest. This is the only example I have seen of the God Displaying the Mask of the God Cocijo in His Headdress. One characteristic of this category is that all the figures are young, sometimes even adolescent, like the figure on this urn. The cult of this deity extended from Monte Albán II through Monte Albán IV and became increasingly complex with elaborate ornamentation and symbolism. The headdress centered around the God Cocijo, here depicted with the characteristic facial mask and "C" glyph and two hands hanging suspended in the lower part of the headdress on each side of the mask. Several of the figures have a second pair of hands in the headdress, as seen on this figure. The faces of the personages on these urns are typically without masks, as portrayed here. There is a trifoliate element and maize symbol,

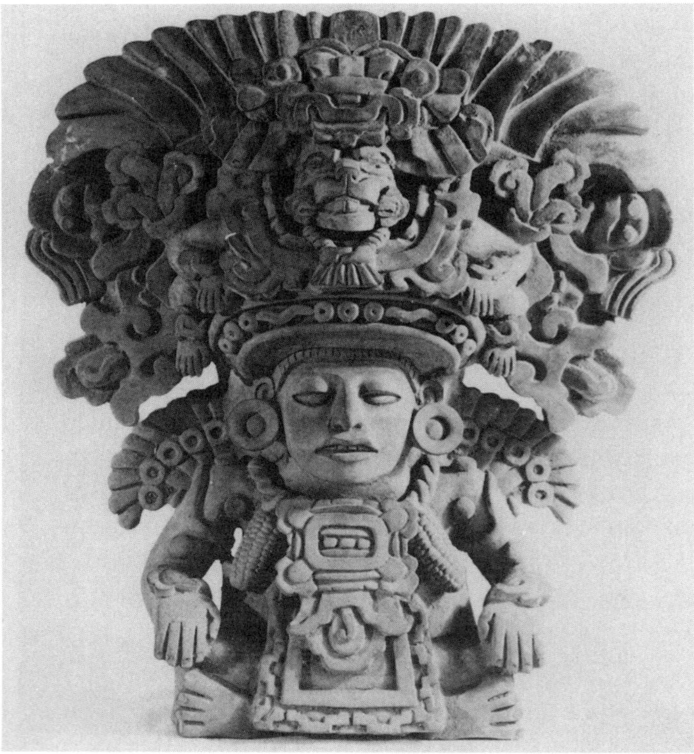

Fig. 110. Z-40. Urn depicting the God displaying the mask of the God Cocijo in his headdress.

also characteristic of the category, on each side of the Cocijo mask in the headdress (Boos 1966b, 65). The naturalistic maize ears hang from the necklace as ornaments on each side of the pectoral glyph, which is unusual. Because morphometric measurements are lacking, it is not possible to identify the maize, but it looks like Nal Tel/Pira. This urn is illustrated by Caso and Bernal (1952, 98–99, fig. 163).

Urns depicting the God with the Ear of Maize in His Headdress

Z-41. Fig. 111. Urn depicting the God with the ear of maize in his headdress. Musée de l'Homme 24.13.3645.
Provenience: Zaachila.
Fabric: gray.
Dimensions: ht. 56.3 cm, width 41.9 cm.
Period: Monte Albán IV.
Race of maize: Niñuelo.

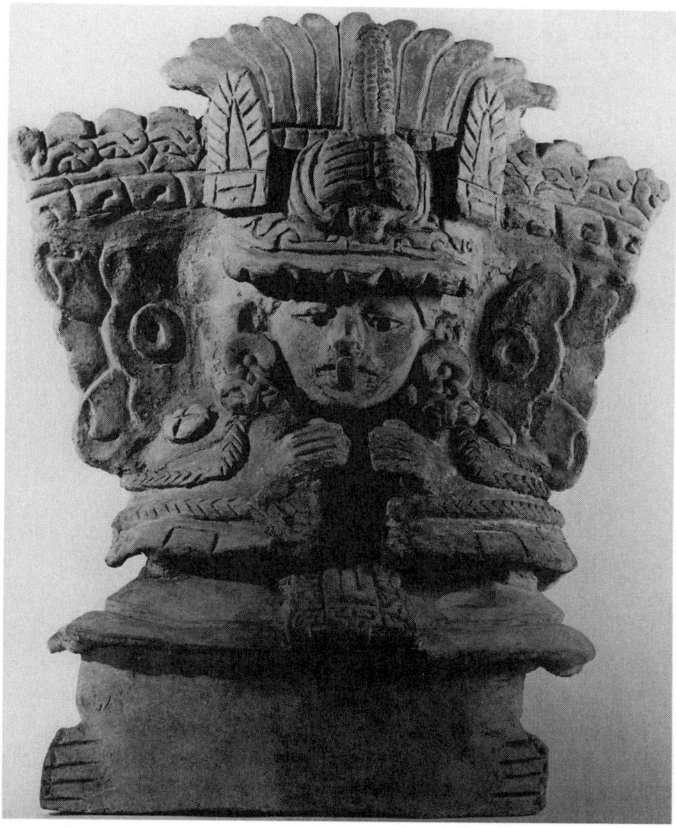

Fig. 111. Z-41. Urn depicting the God with the ear of maize in his headdress.

This urn has been dated by thermoluminescence and confirmed to be a true antiquity. The figure is seated cross-legged with its hands in front. It wears elaborate earspools, a cape, and a loincloth. The central medallion of the headdress is two hands holding an ear of maize flanked by two feather-like elements and surmounted by a penache. In back of the figure is an ornate throne-like structure. Appropriate to the category in accordance with Caso and Bernal (1952), the figure is youthful.

Z-42. Fig. 112. Urn depicting the God with the ear of maize in his headdress.
Museo Frissell de Arte Zapoteca 8011.
Provenience: unknown.
Fabric: dark gray with buff patina.
Dimensions: ht. 16.9 cm, width 10.8 cm.
Period: Monte Albán IV.
Race of maize: Nal Tel/Pira.

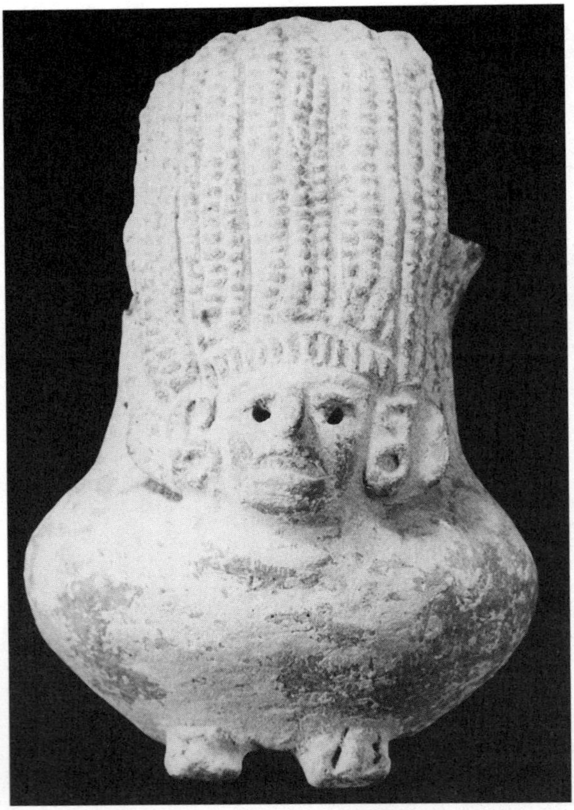

Fig. 112. Z-42. Urn depicting the God with the ear of maize in his headdress.

This jar is in the olla form diagnostic of Monte Albán IV or Mixtec influence. Five flattened standing ears of maize compose the headdress. The face is plain, and the figure wears small plain round earspools. The hands are not depicted, but the feet are. On other ollas, the hands have also been displayed.

Z-43. Fig. 113. Urn depicting the God with the ear of maize in his headdress.
Museo Frissell de Arte Zapoteca 4889.
Provenience: Ejútla.
Fabric: dark brown, highly polished.
Dimensions: ht. 13.5 cm, width 7.5 cm.
Period: Monte Albán IV.
Race of maize: Puya.

This is another unusual vessel that is quite small and a modified olla form resembling a bottle. The clay is hard, highly fired, and smoothly polished.

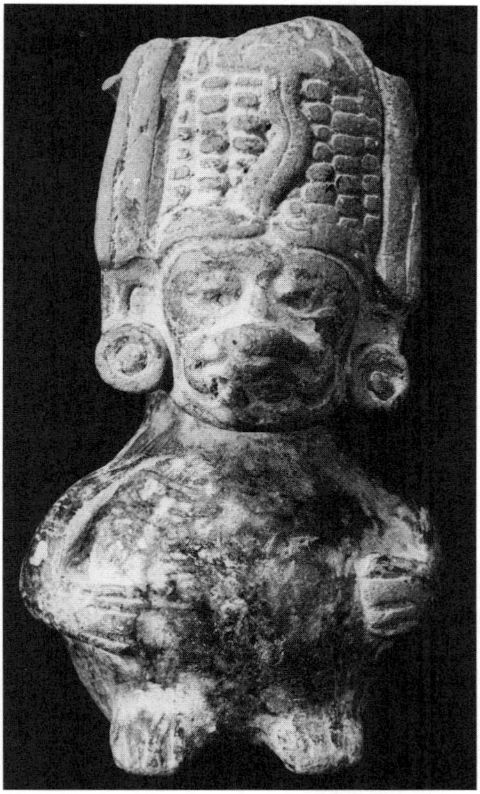

Fig. 113. Z-43. Urn depicting the God with the ear of maize in his headdress.

It is unlike any other Zapotec ceramic piece I have seen. A single ear of maize composes the headdress in turban-like fashion. A curved line extends vertically through the middle of the ear. The face has a mask, but it is damaged and not clearly distinguishable. The figure wears plain round earspools, is standing, and the hands and feet are displayed. Although the race of maize cannot be identified with certainty because the ear has obviously been flattened out and dimensions may be distorted, it seems most closely to resemble Puya.

Z-44. Fig. 114. Urn depicting the God with the ear of maize in his headdress.
Museo Nacional de Antropología 6–6755.
Provenience: unknown.
Fabric: buff.
Dimensions: ht. 34.4 cm, width 24.2 cm.
Period: Monte Albán IV.
Race of maize: Karapampa.

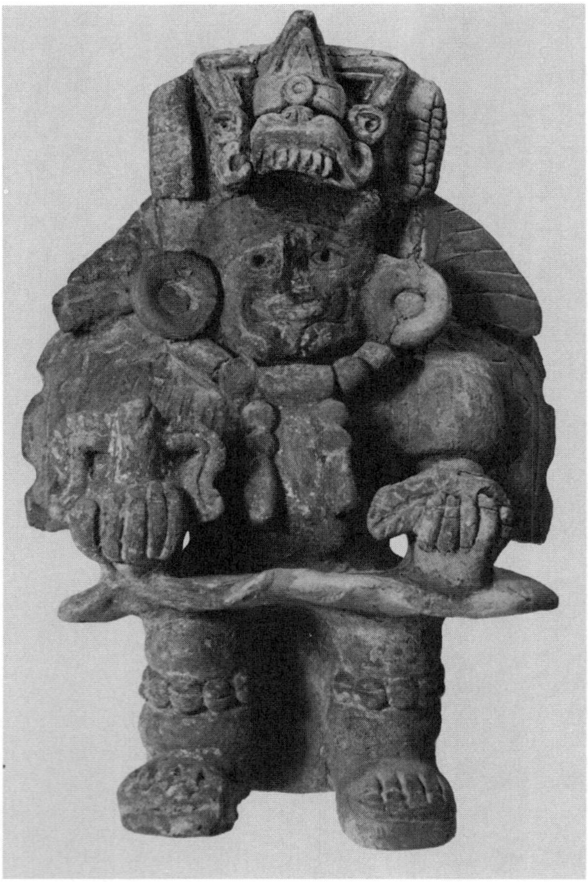

Fig. 114. Z-44. Urn depicting the God with the ear of maize in his headdress.

Although Caso and Bernal (1952, 94, fig. 157) classify this urn as the God of Glyph "L," I have placed it in the category of the God with the Ear of Maize in His Headdress because of the unusual glyph in the headdress. This glyph more closely resembles the year glyph than the "C" glyph (Leigh 1970, 268, figs. 100 and 101; Caso 1928, 47, fig. 22). Boos (1966a, 345, fig. 318) illustrates an urn with a similar element in the headdress and describes it as a mask of the Bat God surmounted by a trapezoid Mixtec Year-Sign.

Urn depicting the God with the Bowknot in His Headdress

Z-45. Fig. 115. Urn depicting the God with the bowknot in his headdress.
Museo Nacional de Antropología 6–6885.
Provenience: Etla.

Fabric: dark gray.
Dimensions: ht. 12.2 cm, width 11.4 cm.
Period: Monte Albán IV.
Race of maize: Chapalote/Nal Tel.

The God with the Bowknot in His Headdress is a member of the "Maize Complex." The main characteristic of the category is the bowknot as the central ornament of the headdress. The figure may appear unmasked or wearing a mask that is characteristic of Cocijo, as it is here, or wearing the mask of the God of Glyph "L" (Boos 1966b, 404). This is the only example in which an urn in this category displays maize with an ear appearing on each side of the vessel. This piece is illustrated by Caso and Bernal (1952, 107, fig. 176).

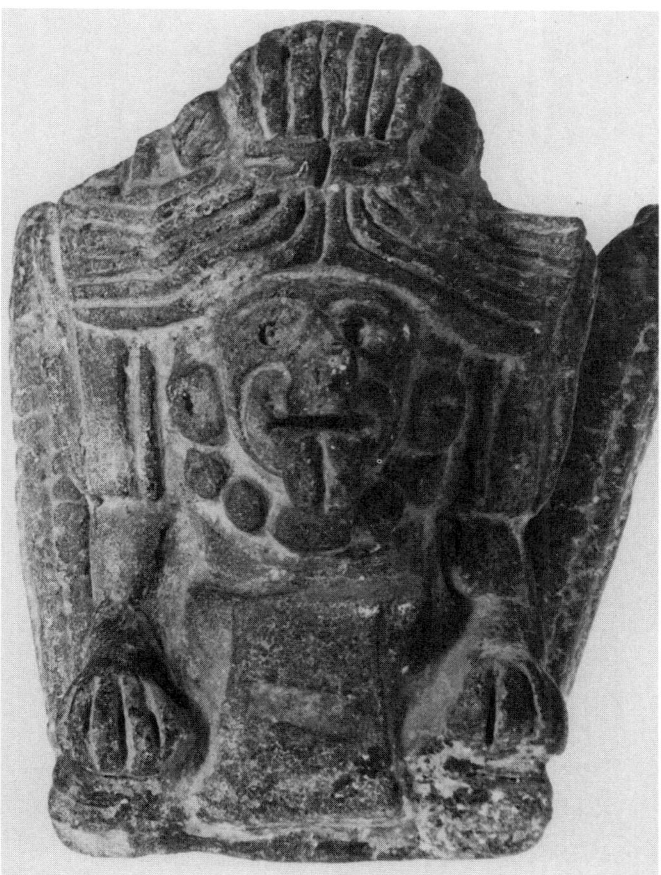

Fig. 115. Z-45. Urn depicting the God with the bowknot in his headdress.

Urn Depicting the Goddess "13 Serpent"

Z-46. Fig. 116. Urn depicting the Goddess "13 Serpent."
Museo Nacional de Antropología 6–6894.
Provenience: Zaachila.
Fabric: gray with red traces.
Dimensions: ht. 14.5 cm, width 14.8 cm.
Period: Monte Albán IV.
Race of maize: Oke.

The attribute that characterizes the Goddess "13 Serpent" is the headdress of woven strands often adorned by ball-shaped ornaments that may represent bells (Boos 1966b, 352). The headdress of this figure shows the woven

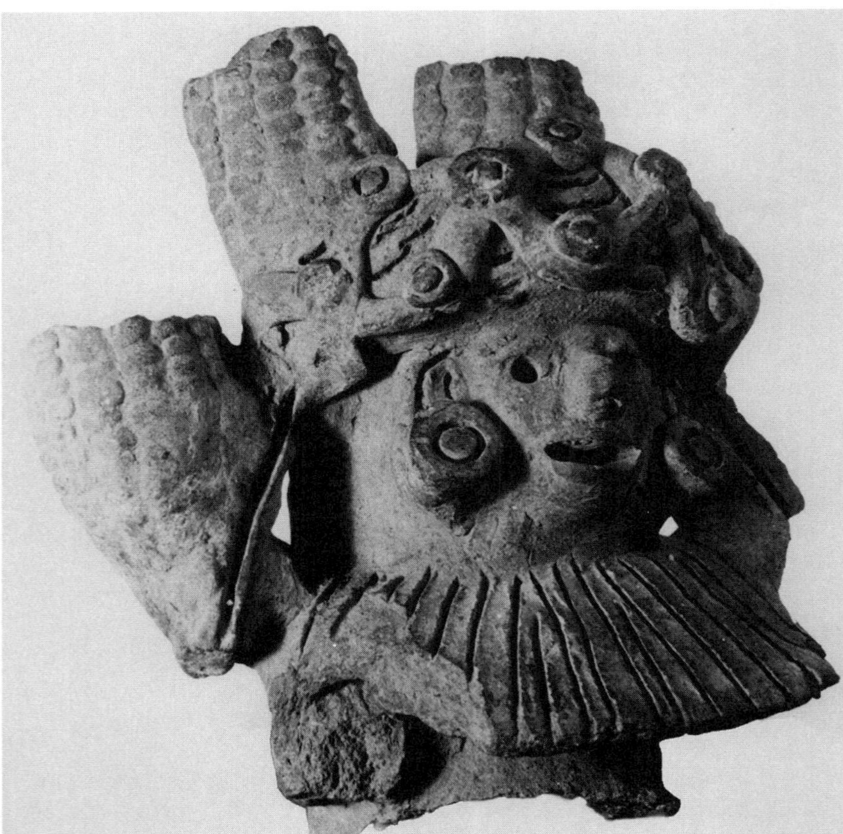

Fig. 116. Z-46. Urn depicting the Goddess "13 Serpent."

strands surmounted by ears of maize. This piece has been damaged and most of the ears of maize are missing so the race of maize cannot be identified. As far as is known, this goddess was not associated with maize, which makes the association here questionable. Also, the agape mouth is not characteristic of Zapotec ceramics.

Urns Depicting the Opossum God

Z-47. Fig. 117. Urn depicting the Opossum God.
Museo Nacional de Antropología 6–965.
Provenience: San Pedro Ixtlahuaca.
Fabric: gray with traces of red.
Dimensions: ht. 28.0 cm.
Period: Monte Albán IV.
Race of maize: unidentified.

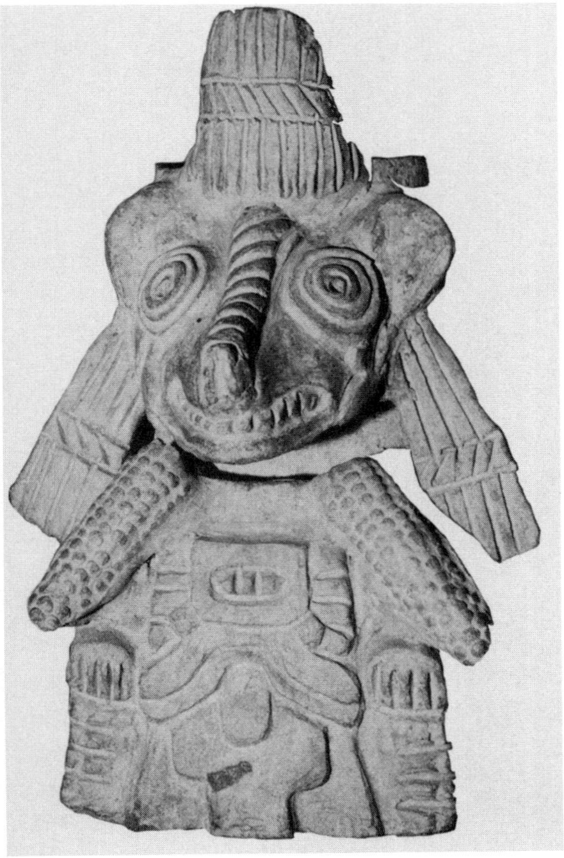

Fig. 117. Z-47. Urn depicting the Opossum God.

Characteristic elements of the Opossum God include round staring eyes that are incised concentric circles, an indented nose ridge that forms a spiral extending up to the forehead, and pulled back lips displaying teeth (Boos 1966b, 237). All these elements are present in this figure. Boos also says that there is a characteristic ornament attached to the cheek or ear, or worn independently, that is leaf-shaped, often showing the leaf veins (Boos 1966b, 239, fig. 217). Some archaeologists believe this represents amaranth. The Opossum God is associated with maize, but the exact nature of the symbolism is unclear. The headdress of this figure forms a bowknot. This urn is illustrated by Caso and Bernal (1952, 272, fig. 423) and Wellhausen et al. (1952, 217, plate B).

Z-48. Fig. 118. Urn depicting the Opossum God.
Museum für Völkerkunde Berlin 29029.

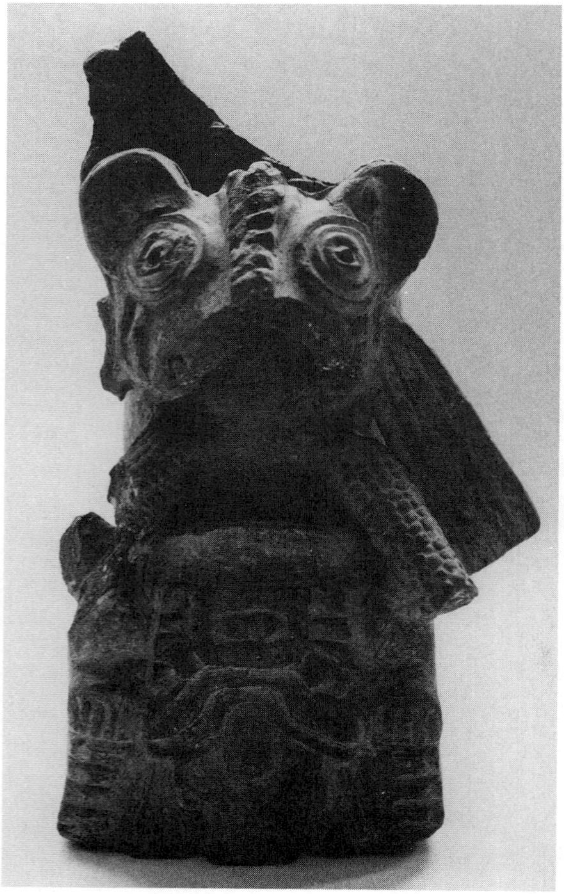

Fig. 118. Z-48. Urn depicting the Opossum God.

Provenience: San Pedro Ixtlahuaca.
Fabric: gray with buff traces.
Dimensions: ht. 30.0 cm, width 18.1 cm.
Period: Monte Albán IV.
Race of maize: Nal Tel/Pira.

In spite of damage, it appears that this urn is a virtual replica of Z-47 with the exception of the headdress, which has a "C" glyph as the central medallion rather than the bowknot. Both urns have the same provenience. The TL date for this piece is A.D. 1057. It is illustrated by Immina Schuler-Schömig (1970, 140, fig. 211).

Unidentified Figures

Z-49. Fig. 119. Urn depicting an unidentified figure.
Museum für Völkerkunde Berlin 28354.
Provenience: Zimatlán.

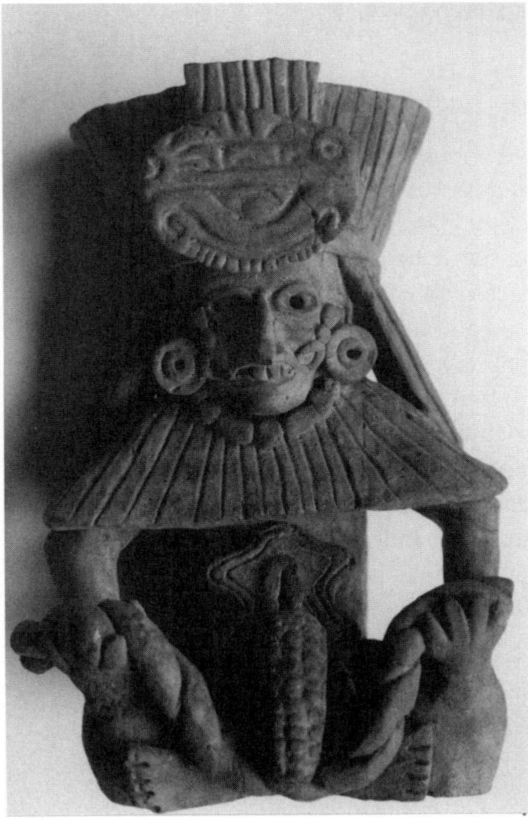

Fig. 119. Z-49. Urn depicting an unidentified figure.

Fabric: gray with buff traces.
Dimensions: ht. 30.0 cm, width 18.1 cm.
Period: Monte Albán V.
Race of maize: Nal Tel/Pira.

This urn is in a category of its own because it has a single ear of mold-made maize as a loincloth. The figure is seated cross-legged with hands resting on the knees and holding a rope that curls down in front of its feet. The face is not masked but does have applied ornamentation on the cheeks. The figure wears a feather cape, round earspools, and a bead necklace. There is a glyph resembling the "C" glyph as the central medallion of the headdress surmounted by a penache that also has flaps projecting down behind the shoulders. The TL date for this piece is A.D. 1208 so it is later than most pieces depicting mold-made maize and is assigned to Monte Albán V.

Z-50. Fig. 120. Aztec brazier depicting an unidentified maize deity.
Yale University Art Museum 1973.88.14.
Provenience: unknown.
Fabric: polychrome white, red, and black.
Dimensions: ht. 45.5 cm.
Period: Postclassic, A.D. 1400–1500.
Race of maize: unidentified.

This complex Aztec brazier is an hourglass-shaped urn open at the top and bottom and has projecting flanges on either side. The rear wall is pierced at midheight by two holes, and the base is ringed by applied overlapping shell-like forms. The standing figure holds a boomerang-shaped object in each hand that may represent ceremonial knives for sacrificial acts. Half of the face represents Tlaloc, as evidenced by the goggle associated with the God of Rain, and the other half is a human face representing the God of Harvest. This figure links the two deities that symbolize related natural phenomena important in agriculture. The figure wears a triangular loincloth, an apron, sandals, earspools, and a bead necklace with five ears of maize attached below. The headdress is striped and composed of four central triangles pointing upward. The conical-shaped maize ears appear to have been impressed from a naturalistic mold but have been retouched to the point of obscuring the data for racial identification. The maize looks like the race Conico.

Z-51. Fig. 3. Urn depicting an unidentified figure (see page 10).
Museo Frissell de Arte Zapoteca 8269.
Provenience: unknown.

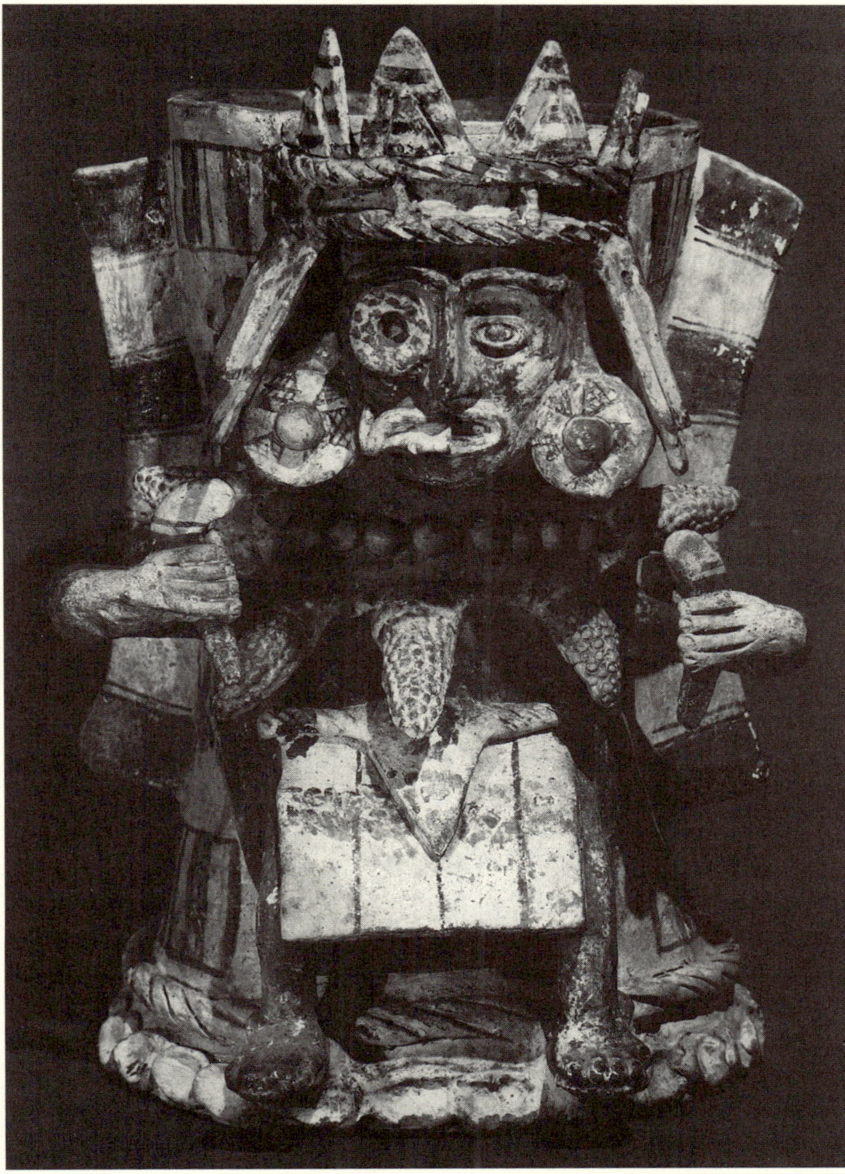

Fig. 120. Z-50. Aztec brazier depicting an unidentified maize deity.

Fabric: highly polished orange with cream patina.
Dimensions: ht. 16.3 cm, width 12.5 cm.
Period: Monte Albán V.
Race of maize: Nal Tel/Pira.

This urn is suspect for several reasons. It has no distinguishing attributes of any established category. It is impossible to tell much about the headdress because most of it has been broken off. The face is unadorned, and the treatment of the mouth with the tongue between the lips is uncharacteristic of Zapotec figures. There appears to be a ruffled collar around the neck, and the loincloth seems to be tied on with a cummerbund. The loincloth's incised design is typical. Although the hands are on the knees, the figure is not seated cross-legged but is in a crouching position not seen in Zapotec ceramics. The placement of the maize on the sides of the vessel is unusual, and the clay is different from any of Zapotec craftsmanship. It has a cream patina that flakes off easily and reveals a bright orange highly polished surface beneath.

Urns Encircled with Ears of Maize

Z-52. Fig. 121. Urn depicting the Goddess "2 J" encircled with ears of maize. Smithsonian Institution National Museum of Natural History 115033.
Provenience: Chicomezuchil.
Fabric: light gray with white traces.
Dimensions: ht. 15.5 cm, width 25.5 cm.
Period: Monte Albán IV.
Race of maize: dent corn.

This vessel is a brazier encircled by maize ears, each impressed from the same mold. It is illustrated by Boos (1966a, 347, fig. 321), who classifies it as the Goddess "2 J." Thermoluminescence dating has established this piece is at least one thousand years old (David Zimmerman personal communication, February 25, 1975). It is the only vessel like this with a figure on it. The main attribute of the Goddess "2 J" is the "J" glyph depicted in the headdress, but that section of the headdress is missing, and the figure has been identified on the basis of the characteristic indigenous feminine dress, the *huipil*, or undergarment, and the *quichquemitl*, the bordered outer garment. In Monte Albán IIIB, this goddess was generally depicted standing with her arms crossed on her bosom or hanging down by her sides (Boos 1966a, 338). On this specimen, however, she is in an unusual kneeling position, but her arms are crossed on her bosom. The kneeling position is rare, and the only reference to figures on urns in this position is by Constantine Rickards (1938, 151). The association of the Goddess "2 J" with maize is not extraordinary because this goddess, along with the Bat God, the God with the Bat God in the Headdress, the God of Glyph "L," and the God with an Ear of Maize as the Central Medallion of the Headdress, compose the five deities of the Zapotec Maize Complex (Caso and Bernal 1952, 65). This urn is illustrated by Wellhausen et al. (1952, 15), who identify the race of maize as dent corn. The identifications from the statistical analysis were Salpor Tardio from Central America and Morado from Bolivia, neither of which is a dent corn. Therefore, I am unable to identify this race, which may be extinct.

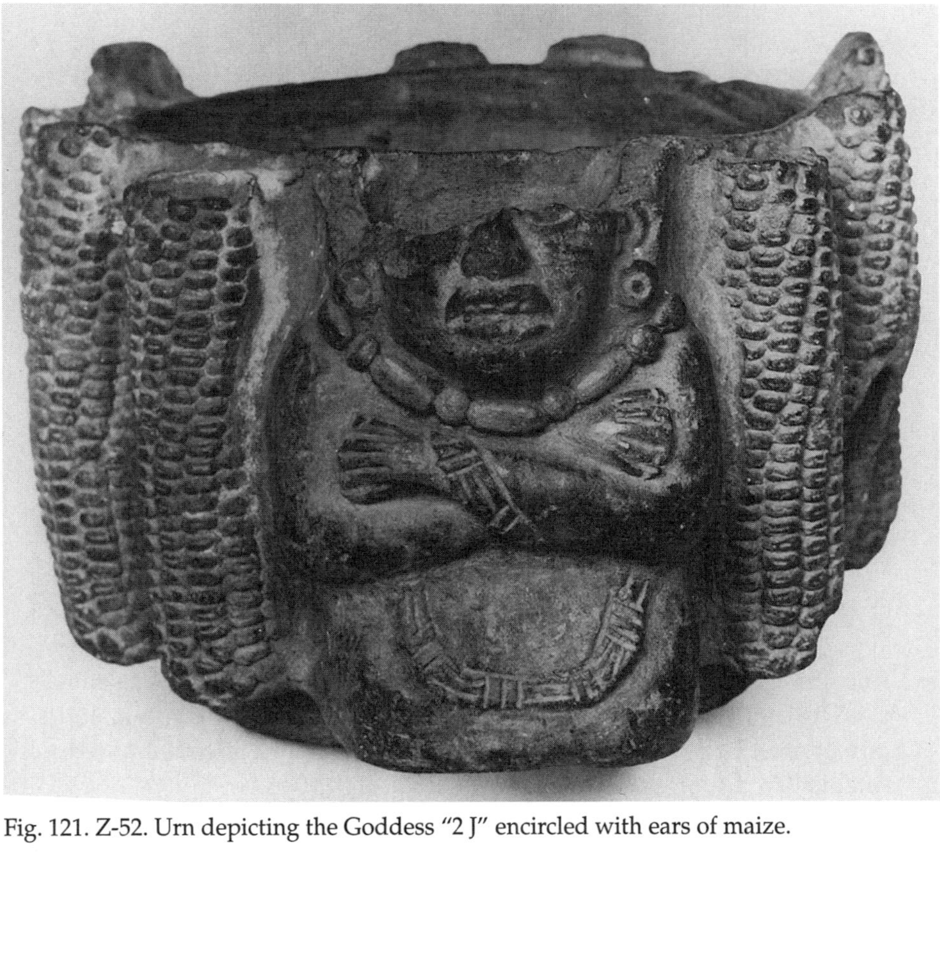

Fig. 121. Z-52. Urn depicting the Goddess "2 J" encircled with ears of maize.

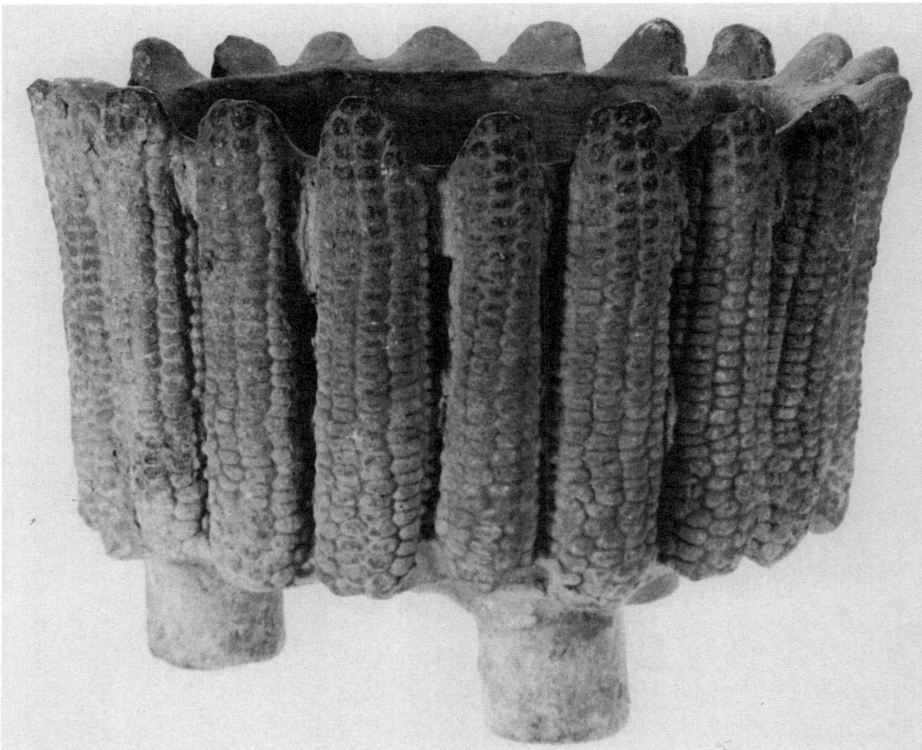

Fig. 122. Z-53. Urn encircled with ears of maize.

Z-53. Fig. 122. Urn encircled with ears of maize.
Museo Nacional de Antropología 6–6012.
Provenience: unknown.
Fabric: light gray.
Dimensions: ht. 15.8 cm, width 21.7 cm.
Period: Monte Albán IV.
Race of maize: Nal Tel/Pira.

This urn is encircled by twenty ears of maize all impressed from the same mold. It is pictured by Wellhausen et al. (1952, 217, fig. C). Of prehistoric corn, it most closely resembles the maize excavated in Cañon del Muerto, Arizona (Anderson and Blanchard 1942, 833–34).

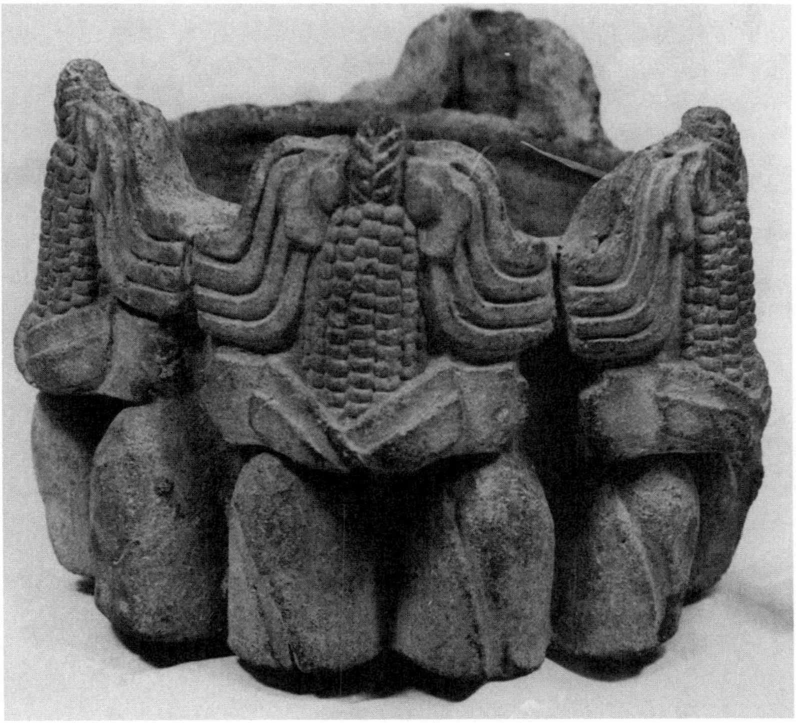

Fig. 123. Z-54. Urn encircled with ears of maize.

Z-54. Fig. 123. Urn encircled with ears of maize.
Tulane University Middle American Research Institute 35–5976.
Provenience: Monte Albán.
Fabric: light gray.
Dimensions: ht. 15.0 cm, width 14.0 cm.
Period: Monte Albán IIIB/IV.
Race of maize: Nal Tel/Pira.

This piece has four ears of maize with staminate tips, trifoliate elements, and stylized silks encircling the vessel. There were originally six maize ears. Beneath the maize, the vessel is encircled by bullet-shaped elements that probably represent bells. This urn was dated by thermoluminescence and found to be at least one thousand years old (David W. Zimmerman personal communication, November 28, 1977).

Mazorca Sherds

Z-55. Fig. 124. Two ears of maize from an urn.
Tulane University Middle American Research Institute 35–6004.
Provenience: Mitla.
Fabric: gray.
Dimensions: ht. 16.4 cm, width 4.0 cm.
Race of maize: Chapalote/Nal Tel.

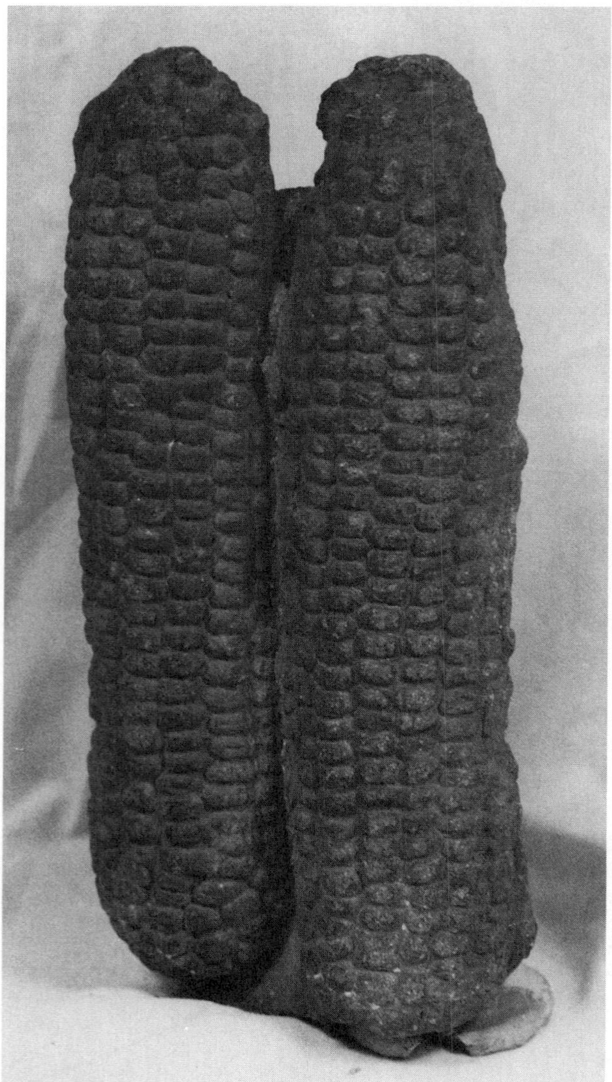

Fig. 124. Z-55. Two ears of maize from an urn.

These mazorcas were dated by thermoluminescence and found to be at least one thousand years old (David W. Zimmerman personal communication, March 28, 1977).

Z-56. Fig. 125. Mazorcas.
Donovan Senter U5/4552.
Provenience: excavated at Zacatepechi near Xochicalco, Morelos.
Fabric: brown.
Dimensions: ht. 8.3 cm, width 3.2 cm.
Period: Early Postclassic, A.D. 900–1100.
Race of maize: Confite Puntiagudo/Canguil.

My attention was drawn to these sherds, which are notable for the distinct imbrication of the kernels and are the only examples of pointed popcorn from Mexico, by Donovan C. Senter (personal communication, November 26, 1975).

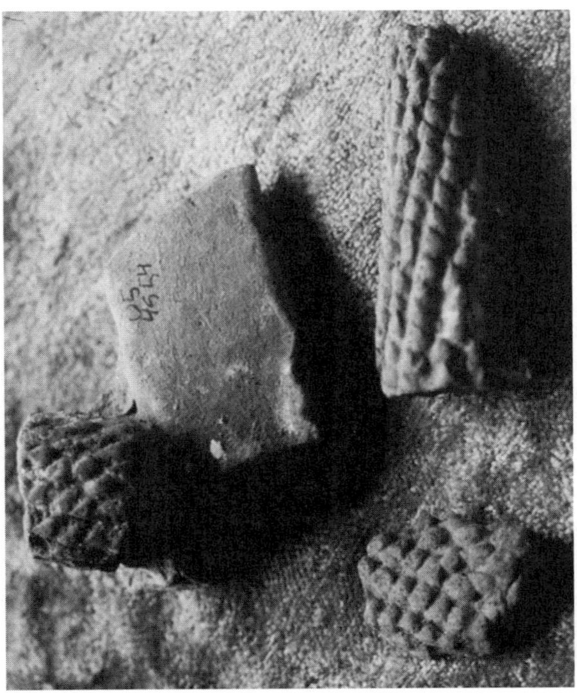

Fig. 125. Z-56. Mazorcas.

4

Corn in the Clay: Races of Maize on Pre-Columbian Pottery

Identifications of maize races depicted on the pottery were made on the basis of information about living races published in the "Races of Maize" monographs (Brieger et al. 1958; Brown 1960; Grant et al. 1963; Grobman et al. 1961; Hatheway 1957; Ramírez et al. 1960; Roberts et al. 1957; Timothy et al. 1961; Timothy et al. 1963; Wellhausen et al. 1952; Wellhausen et al. 1957). Morphometric measurements included row number, ear length, ear diameter, kernel width, and kernel thickness. These data were compared with 249 Latin American maize races in the series of published reports (see Appendix B). All the data were put into a statistical program that chose those of Latin American maize races that most closely matched each pottery replica.

Methodological Considerations

Three factors that should be considered in estimating how reliable the identifications derived from this numerical taxonomic study are population sampling in the original maize collections, comparison of extant maize grown today to ancient specimens, and variability in plant populations caused by environmental differences.

Sampling. Reliability of the identifications on the ceramic models depends on how carefully collected and accurately described the races in the original reports were. The number of collections sampled in the published reports varied from one to fourteen (see table 4.1). The number of ears in each sample was not reported. According to Bird and Goodman (1977:471), "Each collection was usually represented by a sample of 5 to 15 ears, obtained from the same locality and thought by the collector to be of the same type." The data were averaged and the mean values were reported for each trait.

Table 4.1. General characteristics of races of maize depicted on ceramic models

Race (N[a])	Type	Kernel shape	Kernel color	Elevation, m	Country	Days to flowering
Northern South American popcorns[b]						
Nal Tel[c] (1)	Pop	Round	Yellow	100-1800	Mexico	105
Pira[d] (9)	Pop	Round	White	400-2000	Colombia	90
Confite Morocho[e] (3)	Pop	Pointed	Yellow	2500-3000	Peru	116
Confite Puntiagudo[e] (11)	Pop	Pointed	White/red/brown	2500-3500	Peru	125
Canguil[f] (3)	Pop	Pointed	White	2260	Ecuador	-
Imbricado[d] (2)	Pop	Pointed	White	2000-2600	Colombia	146
Chapalote Group[b]						
Chapalote[c] (1)	Pop	Round	Brown	100-600	Mexico	110
Northwestern South American Races[b]						
Pollo[d] (5)	Flint/pop	Round	Yellow	1600-2160	Colombia	143
Chococeño[d] (14)	Flint/pop	Round	White/yellow	0-200	Colombia	101
Chaparreño[e] (1)	Flint/pop	Dent/pointed	White/yellow	10-500	Peru	83
Mochero[e] (6)	Flint/pop	Dent	Red/brown/purple	25-50	Peru	69
South Andean Flints[b]						
Confite Puneño[e] (9)	Flint pop	Round	Purple/red/brown	3600-3900	Peru	91
Karapampa[g] (4)	Flint pop	Round	Orange/red	1800-2160	Bolivia	117
Cuzco Group[b]						
Niñuelo[g] (3)	Flint	Round	White	1900	Bolivia	82
Caribbean Dents[b]						
Puya[h] (3)	Hard dent	Round/dent	Yellow	121	Venezuela	78
Lowland Flours[b]						
Güirua[d] (7)	Flour	Round	Blue	1800-1900	Colombia	84
Not Classified						
Choclero[i] (2)	Flour	Round/pointed	White	58-600	Chile	68
Quicheño Rojo[j] (1)	Flint	Round	Red	1676-2438	Central America	115
Central Andean Complex[b]						
Oke[k] (4)	Flint	Round/pointed	Black	1700-3500	Argentina	

a. N is number of collections from which the measurements reported in the "Races of Maize" publications were made. Number of ears measured was usually 5 to 15. Some measurements were made on ears at the original collection site; others were made on ears of plants grown later at other locations. b. Goodman and Bird 1977. c. Wellhausen et al. 1952. d. Roberts et al. 1957. e. Grobman et al. 1961. f. Timothy et al. 1961. g. Ramírez et al. 1960. h. Grant et al. 1963. i. Timothy et al. 1963. j. Wellhausen et al. 1957. k. Brieger et al. 1958.

Antiquity. Another factor to consider is that maize depicted on the replicas was grown over a millennium ago. How well these modern data correlate with the prehistoric specimens depends on how much or how little change has occurred over fifteen hundred years.

Variability. Because of natural variation within plant populations, characters chosen for use in numerical taxonomy studies like this one should be traits that exhibit minimum variability when grown in different environments. Goodman and Paterniani (1969) examined environmental variation among fifty-five maize races grown in Brazil from 1960 to 1965 and found that reproductive characters, specifically those of the ear, are less susceptible to variation under different environmental conditions than other plant characters. José de Jésus Sánchez-González (1989) studied seventy-one maize races grown in three different environments in Mexico and further demonstrated that ear characters are less subject to environmental variation. Therefore, these are the most reliable characters for numerical taxonomic classification of maize races. Least variable are the characters measured on the ceramic models—ear length, ear diameter, row number, kernel width, kernel thickness. Therefore, when variation occurs in these characters, it is attributable more to genetic constitution than to differences in the environments in which the plants are grown and is more reliable for classification purposes.

Numerical Taxonomy of Ceramic Models

The goal of numerical taxonomy is to use numerical methods to order organisms into groups on the basis of how closely related they are (Sneath and Sokal 1973). Classification is based on a matrix of similarities, then various computational methods are employed to evaluate the taxonomic relationships numerically. This approach is referred to as cluster analysis. Important considerations in numerical taxonomy are the choice of characters, whether the methods for partitioning the variation into components actually reflect biological reality, and adequate sample size.

Mary Eubanks Dunn (1975, 1977, 1979, 1984) measured maize models on pre-Columbian jars and identified the races qualitatively by comparing the archaeological specimens to data and photographs in the "Races of Maize" monographs and to sample ears of selected races provided by Major M. Goodman. Goodman and Bird (1977; Bird and Goodman 1977) employed principal component and unweighted cluster analysis to classify 219 Latin American races of maize into groups. Using the statistical program for clustering maize races, Goodman input data from tables 4, 8, and 9 of Dunn's study (1977) and recovered races that most closely matched the ceramic models. A significant difference between the data sets of Goodman and Bird (1977; Bird and Goodman 1977) and Dunn (1975, 1977, 1997, 1984) was that the former authors included 219 of the 249 maize races in the published reports omitting Central America and made some changes in the published measurements, whereas Dunn included all 249 races and followed measurements as originally reported.

Goodman ran two different statistical programs comparing the ceramic morphometric data with data from 219 races of maize. The first program, called DUNNST, standardized the data by subtracting overall means from the 219 races and dividing by overall standard errors in the 219 races to calculate Euclidean distances between each of the ceramic specimens. It then ordered the distances from each race to the specimens, using a simple unweighted averaging method to cluster the ceramic models together. The second program, called DUNNTR, was similar but used a complex transformation of the ceramic data to account for mutual correlations among characters. The results of DUNNTR are more conservative "but are more difficult to calculate and explain" (Goodman personal communication, March, 15, 1997). Goodman provided Dunn with the computer printouts for both programs, and the results were considered in making identifications. Because most of the answers did not resemble the ceramic models, however, Dunn decided to identify the races using a qualitative approach. Table 4.2 reports racial identifications with the closest distance to the 219 maize races computed by both of Goodman's statistical programs (Goodman personal communication, March, 15, 1997) and compares them with identifications reported by Dunn (1977). Most of the identifications made by Dunn (1977), computed by Goodman (1977 personal communication), and published by Goodman (1994) are different from the ones reported here. The races identified in those studies that match identifications reported here appear with a footnote b in table 4.2.

Numerical Taxonomy Revisited

Since Dunn's original study and Goodman's statistical analysis, the sample size of prehistoric maize jars has more than doubled, and new information concerning variability in specific traits resulting from environmental effects has been forthcoming (Sánchez G. 1989; Sánchez G. and Goodman 1992). Sánchez-González (1989) showed that there is minimal variation in the ear characters that are employed in this study when grown in different environments. Therefore, expanded sample size and the new findings led

Table 4.2. **Maize races in Goodman's statistical analyses compared to Dunn's identifications**[a]

Specimen	Goodman Programs: DUNNST	Goodman Programs: DUNNTR	Dunn 1977
Frissell 4279 (Z-34)	Pollo	Pollo	Nal Tel[b]
MAI 23/196 (P-3)	Pollo	Nal Tel[b]	Confite Morocho
Larco 2300 (P-24)	Pollo[b]	Pollo[b]	Pollo[b]
Larco 2295 (P-10)	Pollo	Pollo	Kcello
Brüning 7761 (P-65)	Pollo	Pollo	Huancavelicano
Brüning MB1918 (P-66)	Pollo[b]	Pollo[b]	Huancavelicano
Brüning B-99 (P-45)	Pollo	Nal Tel	Confite Morocho[b]
Larco S-39160 (P-62)	Huancavelicano	Huancavelicano	Güirua
Larco 3993 (P-29)	Huancavelicano	Kcello	Guaribero
MMA 67.167.19 (P-58)	Huancavelicano	Pollo	Pollo
Brüning MB1927a (P-68)	Huancavelicano	Kcello	Huancavelicano
Brüning MB1927b (P-68)	Huancavelicano	Pollo	Huancavelicano
MNAA 1/1680 (P-60)	Huancavelicano	Huancavelicano	Güirua[b]
MNAA 36/1376 (P-69)	Huancavelicano	Pollo	Huancavelicano
Larco J-432 (P-5)	Kcello	Kcello	Pollo
MAI 5/1741 (P-37)	Kcello	Pollo	Pira
Duke 73.1.408 (P-70)	Kcello	Kcello	Pagaladroga
Brooklyn 1890 (P-22)	Kcello	Pollo	Mochero
MAI 7/2734 (P-71)	Kcello	Pollo	Kculli
Larco 3993 (P-29)	Kcello	Kcello	Guaribero
Larco 2281 (P-39)	Kcello	Kcello	Huancavelicano
Amano 59 (P-14)	Kcello	Kcello	Kcello
MNAA 1/2609 (P-64)	Cabuya	Cabuya	Imbricado
UM NA2141 (Z-16)	Cabuya	Cabuya	Chapalote
MNA 6-6885 (Z-45)	Cabuya	Cabuya	Chapalote[b]
Frissell 8269 (Z-51)	Cabuya	Kcello	Chapalote
UM 67-36-1 (P-36)	Cabuya	Kcello	Pollo
MNAA 1/027 (P-7)	Cabuya	Cabuya	Clavo
MNAA 1/1681 (P-59)	Cabuya	Cabuya	Perlilla
Brüning B-1917 (P-67)	Cabuya	Cabuya	Pollo
MNAA KK/16591 (P-8)	Araucano	Araucano	Mochero
MAI 4/8898 (P-43)	Araucano	Araucano	Pira[b]
MNAA KK/16624 (P-21)	Araucano	Pollo	Mochero
MNAA KK/16583 (P-1)	Araucano	Araucano	Karapampa
MNAA 158 (P-2)	Araucano	Araucano	Karapampa
MNA 6-6012 (Z-53)	Huandango	Huandango	Dzit-Bacal
MARI 35-6004 (Z-56)	Huandango	Huandango	Güirua

(continued)

Specimen	Goodman Programs DUNNST	DUNNTR	Dunn 1977
Larco 4677 (P-61)	Sabanero	Sabanero	Güirua
MNAA KK/16602 (P-54)	Cuzco Gigante	Cuzco Gigante	Perla
Duke 73.1.507 (P-51)	Cuzco Gigante	Chillo	Perla
MNAA 1/2865 (P-32)	Cuzco	Cuzco	Chococeño
Brooklyn 41.1275 (P-25)	Patillo	Pollo	Enano
Larco J-30839a (P-26)	Patillo	Pollo	Polulo
MNAA 2/4756a (P-9)	Clavito	Araucano	Polulo
MNAA 2/4756b (P-9)	Confite Puneño	Pollo	Polulo
Larco J-30839b (P-26)	Confite Puneño	Confite Puneño	Polulo
MNA 6-6708 (Z-28)	Niñuelo	Niñuelo	Nal Tel[b]

a. Goodman (personal communication 1977); Dunn 1977.
b. This identification is the same one reported in this study.

to reconsideration of the application of numerical taxonomy to this data set in order to resolve discrepancies in the results reported earlier by Goodman and Dunn and to improve identifications.

A new statistical program was devised in which characters were weighted according to degree of intraracial genetic variation expected for each trait. This is a significant departure from Goodman's approach, which used overall means for 219 races divided by overall standard errors and unweighted cluster analysis.

Expected variation per trait was initially estimated by measuring row number, ear length, ear diameter, and width and thickness of the kernels on four to thirteen ears of fifteen different maize races: Harinoso de Ocho (N = 6), Chapalote (N = 6), Karapampa (N = 4), Nal Tel (N = 6), Lenha (N = 6), Conico (N = 6), Chake-Sara (N = 4), Canguil Grueso (N = 5), Guaribero (N = 5), Mochero (N = 6), Argentine Pop (N = 6), Enano (N = 7), Pisankalla (N = 4), Reventador (N = 6), and B73 (N = 13). The mean and standard deviations were calculated for each trait. The degree of variation per trait was consistent across all races. The least variation was in kernel width and the greatest in ear length, findings consistent with Sánchez-González's study (1989). The standard deviations estimated for each trait, in each population, were then combined to give a single standard deviation estimate for a particular trait. Ideally, we would like a separate standard deviation for every trait-race combination, but such large-scale data are not available. In the statistical program ear characters were weighted in accordance with their average standard deviations; traits with greater intraracial variation were downweighted; and less variable traits were given more weight corresponding to degree of expected variation. To address the question of how well numerical taxonomy can classify a biological specimen, a blind test was run on specimens of known races. Initially, 10 percent of the answers obtained were correct. The program was subsequently tested and the trait specific standard deviation estimates adjusted to maximize number of correct identifications based on the trial set of known specimens. The highest accuracy that could be attained was that 50 percent of the time a correct identification appeared among the top ten races provided by the statistical program. This program was precisely tailored to give the best possible answer that could be derived using the quantitative method. It went beyond the more general, unweighted cluster analysis by Goodman and Bird (1977; Bird and Goodman 1977) because the program was tested and degree of accuracy defined. The findings indicate that numerical taxonomy is useful in delineating possible taxonomic choices, but greatest accuracy requires careful attention to qualitative differences that distinguish the various maize types. Therefore, in this study quantitative and qualitative methods were combined. Possible identifications were found initially among the top ten choices provided by the statistical analysis; then the pottery specimen was compared with photographs or specimens of those races, along with other information provided in the "Races of Maize" monographs. The race that the ceramic model most closely resembled was selected as the best identification.

Another issue that was addressed using statistics was the effect of clay shrinkage on accuracy of identification. Experiments using real ears to replicate how prehistoric potters made these molds were carried out. Clay composition approximated the terra-cotta from which the archaeological pots were made. Ears were pressed into slabs of clay, then removed when the clay had dried to form a negative cast; the mold was fired, then clay was pressed into the mold; casts were removed when dry and fired. Measurements of ear diameter, ear length, kernel width, and kernel thickness were made on the maize ear, the fired mold, and the fired replica. Shrinkage was approximately 7 percent in each step of this two-step manufacturing process. This is only an estimate because shrinkage can vary depending on composition of the clay and firing temperatures (Donnan 1992). Prehistoric clays were often coarser and firing temperatures may not have been as high as experimental temperatures in a modern kiln. Therefore, conservatively to reflect actual size of the maize ears before shrinkage caused by the manufacturing process, programs were run in which morphometric measurements on all the pottery specimens were increased by 10 percent and 15 percent. The answers generated for the ceramic models were the same when the data were normalized by 10 percent and 15 percent as they were when not inflated. This indicates that shrinkage is not a confounding factor in the identification of maize races depicted on ceramic models. Therefore, the morphometric measurements were not adjusted to accommodate shrinkage.

The Races of Maize

Sixteen races of maize from ten Latin American countries were identified on 129 ceramic models. There are fourteen races on seventy-two Peruvian jars, and eight on fifty-six Oaxacan specimens. Six races appear on jars from both regions. Table 4.3 compares the data for the ceramic specimens from Peru with measurements for the races identified, and table 4.4 gives ceramic specimens from Mexico and their botanical identifications. In four cases, two choices were equally plausible. All possible choices are listed in tables 4.1, 4.3, and 4.4 and are described below. That these specific races are consistently linked as choices in every case in which they are identified suggests that they are closely related and may be the same races of maize but were originally assigned different names because they were collected in different countries by different researchers. Figure 126 shows the geographical distribution of maize races identified on pre-Columbian pottery.

Goodman (1994) pointed out that a sample size of five or more specimens gives high probability of accurate racial identification in maize, whereas misidentification is more frequent when there are fewer than five specimens. Therefore, it is highly probable that identifications of races for which there are five or more specimens are correct. These include sixty-five of Pira/Nal Tel (twenty-nine from Peru and thirty-six from Oaxaca), thirteen of Confite Morocho (all from Peru), seven of Chapalote (four from Peru and three from Oaxaca), seven of Karapampa (four from Peru and three from Oaxaca), and six of Confite Puñeno/Quicheño Rojo (all from Peru). Although those with fewer than five samples may not represent the exact race identified, they closely approximate the type of maize depicted. These include four of Confite Puntiagudo/Canguil (all from Mexico), four of Oke (three from Peru and one from Oaxaca), three of Choclero (all from Peru), two of Chaparreño (all from Peru), two of Mochero (one from Peru and one from Oaxaca), two of Niñuelo (one from Peru and one from Oaxaca), two of Pollo (all from Peru), two of Imbricado (Peru), one of Chococeño (Peru), one of Güirua (Peru), and one of Puya (Oaxaca). The fact that Nal Tel, Confite Morocho, Confite Puneño, Pollo, Confite Puntiagudo, and Chaparreño have been identified among prehistoric maize remains as

Table 4.3. Races of maize identified on Peruvian jars

				Kernel characters	
	Length, cm	Diam, cm	Row no.	Width, mm	Thickness, mm
Race					
Pira (Venezuela)[a]	10.9	2.6	10.8	5.8	3.0
Pira (Colombia)[b]	11.2	2.5	11.0	6.5	3.6
Nal Tel (Mexico)[c]	7.9	2.7	11.4	6.7	3.9
Specimen					
MAI 23/196 (P-3)	8.8	2.3	10.0	6.0	3.5
Larco J-432 (P-5, Fig. 12)	8.9	3.0	10.0	6.3	3.4
MNAA KK/16591 (P-8, Figs. 15, 16)	8.5	2.5	10.0	6.4	4.0
MNAA 2/4756 (P-9, Figs. 17, 18)	11.4	2.5	10.0	5.8	4.1
MNAA 2/4756 (P-9, Figs. 17, 18)	4.7	2.3	10.0	5.8	4.1
Larco 2995 (P-13, Fig. 23)	7.5	2.7	8.0	7.5	3.8
AMNH B/8863 (P-19, Fig. 29)	-	2.3	8.0	7.5	4.0
AMNH 41.2/7427 (P-20, Fig. 30)	8.8	2.0	8.0	7.0	3.8
MNAA KK/16624 (P-21, Figs. 31, 32)	7.5	2.2	8.0	6.0	3.2
Brooklyn 1890 (P-22, Fig. 33)	8.0	2.7	10.0	6.0	2.5
Berlin 17445 (P-23)	4.8	2.0	8.0	4.0	4.7
Berlin 17445 (P-23)	5.2	2.5	8.0	5.9	3.0
AMNH B8978 (P-28)	9.3	3.2	10.0	7.0	3.8
Larco 3993 (P-29, Fig. 40)	8.4	3.0	10.0	6.4	3.7
Yale Peabody 235134 (P-30, Fig. 41)	6.0	2.5	10.0	4.0	1.9
Yale Peabody 235134 (P-30, Fig. 41)	4.4	3.0	10.0	7.0	3.5
AMNH 41.2/7431 (P-31, Fig. 42)	8.8	2.0	8.0	7.0	3.8
Paris 87.134.14 (P-33, Fig. 45)	6.3	2.8	10.0	5.7	2.0
FMNH 1229 (P-34, Fig. 46)	7.8	2.6	8.0	7.5	3.7

FMNH 4794 (P-35, Fig. 47)	6.0	2.5	10.0	5.0	3.3
Larco 2281 (P-39, Fig. 50)	9.0	2.9	8.0	7.0	4.0
Berlin 17826 (P-40, Fig. 51)	6.3	2.9	12.0	5.9	3.7
Berlin 17826 (P-40, Fig. 51)	4.5	2.6	8.0	7.5	3.8
Berlin 17826 (P-40, Fig. 51)	5.9	2.9	10.0	5.5	3.0
FMNH 4796 (P-41, Fig. 52)	8.0	3.0	10.0	6.3	3.5
FMNM 5667 (P-42, Fig. 53)	5.4	2.5	10.0	6.8	3.5
FMNH 5667 (P-42, Fig. 53)	8.2	2.4	12.0	6.3	3.1
MAI 4/8898 (P-43)	8.8	2.2	10.0	6.0	3.5
AMNH B/5007 (P-44, Fig. 54)	—	2.3	8.0	6.9	3.5
Berlin 17820 (P-49, Fig. 61)	6.0	2.2	10.0	5.0	3.1
BM 1921.10-27172 (P-52, Fig. 63)	8.1	2.7	8.0	7.5	3.7
Berlin 17836 (P-62, Fig. 70)	8.2	2.5	12.0	5.0	3.1
Duke 73.1.408 (P-69, Fig. 77)	8.6	2.8	10.0	7.0	3.8
Berlin 3571 (P-70, Fig. 78)	4.7	2.3	8.0	6.4	3.1
MAI 7/2734 (P-71, Fig. 79)	7.7	2.3	10.0	6.5	3.6
BM 1921.10.27 (P-72, Fig. 80)	8.6	2.6	8.0	6.7	4.0

Race					
Chapalote (Mexico)[c]	11.0	2.9	12.3	6.7	4.1
Nal Tel (Mexico)[c]	7.9	2.7	11.4	6.7	3.9
Specimen					
UM 67-36-1 (P-36, Fig. 48)	10.4	2.6	10.0	7.0	3.7
Paris 83.30.161 (P-48, Figs. 59, 60)	6.5	2.4	8.0	8.5	4.3

(continued)

Table 4.3.—*continued*

	Length, cm	Diam, cm	Row no.	Kernel characters	
				Width, mm	Thickness, mm
Specimen					
MMA 67.167.19 (P-57, Fig. 67)	6.2	2.7	10.0	6.5	4.0
MNAA 1/1681 (P-58, Fig. 68)	11.9	2.7	12.0	6.0	4.9
MNAA 1/1681 (P-58, Fig. 68)	10.8	3.5	14.0	6.4	4.5

Race					
Confite Morocho (Peru)[d]	6.7	2.3	10.6	5.9	4.6
Specimen					
MNAA KK/16583 (P-1, Fig. 10)	8.2	2.0	8.0	6.3	4.0
MNAA 158 (P-2, Fig. 11)	8.2	2.0	8.0	7.5	4.0
Paris 78.2/752 (P-6, Fig. 13)	6.0	2.1	10.0	6.6	4.2
Larco 2295 (P-10, Fig. 19)	6.5	2.4	8.0	7.0	3.5
Berlin 17819 (P-17, Fig. 27)	5.4	2.7	8.0	6.5	4.6
Berlin 17819 (P-17, Fig. 27)	—	2.6	10.0	7.2	4.7
FMNH 4688 (P-18, Fig. 28)	8.0	2.2	8.0	6.8	3.8
Brooklyn 41.1275-74 (P-25, Fig. 36)	6.6	2.8	12.0	5.0	4.8
Larco J-30839 (P-26, Figs. 37, 38)	6.5	2.8	14.0	5.0	4.7
Larco J-30839 (P-26, Figs. 37, 38)	3.5	2.0	10.0	5.0	4.3
MAI 5/1741 (P-37, Fig. 49)	7.5	2.4	10.0	6.0	3.6
Berlin 17827 (P-38)	6.0	2.0	8.0	6.4	3.7
Brüning B-99 (P-45, Figs. 55, 56)	8.5	1.5	8.0	6.2	3.6
Brüning B-99 (P-45, Figs. 55, 56)	3.3	1.4	8.0	6.0	2.5
Brüning 7761 (P-64, Fig. 72)	5.6	1.8	6.0	6.7	3.5
Brüning B-1918 (P-65, Fig. 73)	5.6	1.8	6.0	6.7	3.5

Race					
Confite Puneño (Peru)[d]	6.6	3.5	16.0	6.7	5.0
Quicheño Rojo (Central America)[e]	13.1	3.4	13.0	7.0	4.8
Specimen					
Paris 78.2.94 (P-11, Fig. 20)	8.5	3.5	12.0	6.6	4.4
Linden 69037 (P-12, Figs. 21, 22)	8.0	3.6	10.0	6.5	4.1
Berlin 3727 (P-15, Fig. 25)	10.6	3.9	12.0	6.1	6.0
Berlin 17817 (P-16, Fig. 26)	8.8	3.3	12.0	6.2	5.4
FMNH 5667 (P-42, Fig. 53)	10.0	3.0	12.0	5.8	5.0
MNAA 1/1680 (P-59)	13.8	3.8	14.0	6.5	4.6

Race					
Karapampa (Bolivia)[f]	12.2	3.0	8.8	7.9	3.3
Specimen					
MRH AAM.39.107 (P-47, Fig. 58)	6.1	2.5	8.0	8.1	3.8
Brüning MB-1927 (P-67, Fig. 75)	7.5	2.7	8.0	7.5	3.5
Brüning MB-1927 (P-67, Fig. 75)	5.9	2.4	8.0	7.5	3.5
MNAA 36/1376 (P-68, Fig. 76)	6.0	2.6	8.0	8.8	3.0

Race					
Oke (Brazil)[g]	8.2	3.4	10.7	7.9	4.3
Specimen					
MNAA 1/027 (P-7, Fig. 14)	10.2	3.4	10.0	7.5	4.7
Linden 69037 (P-12, Figs. 21, 22)	7.3	3.1	8.0	8.2	4.4
Linden 69037 (P-12, Figs. 21, 22)	8.0	3.5	8.0	8.3	4.3
Amano 59 (P-14, Fig. 24)	8.2	3.5	8.0	8.0	3.9

(*continued*)

Table 4.3.—*continued*

	Length, cm	Diam, cm	Row no.	Kernel characters	
				Width, mm	Thickness, mm
Race					
Choclero (Chile)[h]	14.9	5.6	22.1	7.1	4.1
Specimen					
MNAA KK/16602 (P-53, Fig. 64)	13.5	5.5	22.0	7.5	4.5
Duke 73.1.507 (P-54, Fig. 65)	12.3	5.7	22.0	6.3	5.6
Berlin 2797 (P-55)	13.2	6.0	18.0	6.0	5.3
Race					
Chaparreño (Peru)[d]	10.5	4.2	12.0	8.4	4.4
Specimen					
Berlin 4652 (P-27, Fig. 39)	9.9	4.1	12.0	8.4	4.1
Berlin 18175 (P-56, Fig. 66)	11.7	3.8	16.0	7.9	4.5
Race					
Mochero (Peru)[d]	7.6	4.1	14.0	7.6	4.5
Specimen					
Larco S-39160 (P-61, Fig. 69)	14.0	3.5	14.0	7.0	4.3
Larco S-39160 (P-61, Fig. 69)	10.3	3.9	14.0	7.0	4.4
Race					
Pollo (Colombia)[b]	8.8	2.9	10.0	7.9	5.0
Specimen					
Larco 2300 (P-24, Figs. 34, 35)	7.0	2.3	8.0	7.0	4.6

Brüning B-1917 (P-66, Fig. 74)	8.0	2.2	8.0	8.3	4.2
Race					
Niñuelo (Bolivia)[f]	12.1	3.2	8.2	9.5	4.3
Specimen					
Berlin 3304 (P-51, Fig. 62)	9.0	3.1	10.0	9.1	4.0
Race					
Güirua (Colombia)[b]	16.2	3.4	12.2	8.0	4.9
Specimen					
Larco 4677 (P-60)	12.9	4.5	10.0	7.5	4.8
Race					
Imbricado (Colombia)[b]	11.0	3.9	14.3	7.2	5.3
Specimen					
MNAA 1/2609 (P-63, Fig. 71)	14.6	3.7	12.0	6.2	5.5
Race					
Chococeño (Colombia)[b]	11.0	3.7	17.2	6.5	4.1
Specimen					
MNAA 1/2865 (P-32, Figs. 43, 44)	11.0	4.0	16.0	6.0	3.1

a. Grant et al. 1963.
b. Roberts et al. 1957.
c. Wellhausen et al. 1952.
d. Grobman et al. 1961.
e. Wellhausen et al. 1957.
f. Ramírez et al. 1960.
g. Brieger et al. 1958.
h. Timothy et al. 1961.

Table 4.4. Races of maize identified on Mexican specimens

	Length, cm	Diam, cm	Row no.	Kernel characters	
				Width, mm	Thickness, mm
Race					
Nal Tel (Mexico)[a]	7.9	2.7	11.4	6.7	3.9
Pira (Colombia)[b]	11.2	2.5	11.0	6.5	3.6
Specimen					
MNA 6-2293 (Z-2, Fig. 84)	—	2.3	10.0	4.8	3.6
MNA 6-2290 (Z-3)	—	2.3	10.0	4.8	3.6
MNA 6-2289 (Z-4)	—	2.2	10.0	4.5	3.6
MNA 6-6717 (Z-5, Fig. 85)	—	2.5	8.0	7.0	4.6
MNA 6-6754 (Z-6, Fig. 86)	—	2.2	8.0	6.5	3.8
Frissell 6883 (Z-7, Fig. 87)	—	2.7	10.0	7.2	3.6
MNA 6-338 (Z-8, Fig. 88)	—	2.5	12.0	6.0	4.0
Frissell 8870 (Z-10, Fig. 90)	—	2.6	10.0	7.5	4.2
Frissell S-1796 (Z-11)	—	2.8	10.0	7.5	4.2
MNA 6-5421 (Z-12, Fig. 91)	—	2.6	8.0	6.8	4.6
MNA 6-6757 (Z-13, Fig. 92)	—	1.9	8.0	5.3	4.0
Yale Art Gallery 1973.88.10(Z-14, Fig. 93)	—	2.7	8.0	7.0	3.8
MNA 6-6736 (Z-15, Fig. 94)	—	2.4	8.0	5.0	3.7
MNA 6-275 (Z-17, Fig. 96)	—	2.1	8.0	6.2	4.0
MNA 6-328 (Z-18)	—	2.3	10.0	6.0	3.9

St. Louis 68:452 (Z-19, Fig. 97)	—	2.5	8.0	6.2	4.1
St. Louis 3051.73 (Z-20)	—	2.5	8.0	6.2	4.0
Munich 10.3505 (Z-21)	—	2.3	8.0	6.4	3.5
Munich 14.8.6 (Z-22)	—	2.5	8.0	6.8	3.6
Munich 10.3506 (Z-23)	—	1.6	8.0	5.0	3.3
AMNH 30.2/6099 (Z-25, Fig. 99)	—	2.5	8.0	7.5	4.0
MNA 6-6708 (Z-28, Figs. 102, 103)	7.6	2.3	10.0	5.0	3.0
MNA 6-6758 (Z-29, Fig. 104)	—	2.0	8.0	5.5	3.6
MNA 6-6672 (Z-31, Fig. 106)	—	2.5	10.0	6.0	4.0
MNA 6-6674 (Z-32)	—	2.3	10.0	6.0	3.8
MNA 6-6697 (Z-33)	—	2.6	10.0	6.0	3.6
Frissell 4279 (Z-34, Fig. 107)	5.9	1.9	8.0	5.8	3.5
Munich 04.95 (Z-35)	—	1.8	6.0	7.5	4.2
AMNH 30/95 (Z-38)	—	3.7	12.0	6.2	5.5
AMNH 30/11173 (Z-39, Fig. 109)	—	2.8	10.0	6.6	4.0
Frissell 8011 (Z-42, Fig. 112)	—	1.8	8.0	5.0	3.6
Berlin 29029 (Z-48, Fig. 118)	7.5	2.5	10.0	6.3	4.5
Berlin 28354 (Z-49, Fig. 119)	4.5	2.0	8.0	5.0	3.6
Frissell 8269 (Z-51, Fig. 3)	10.0	2.7	12.0	6.0	3.8
MNA 6-6012 (Z-53, Fig. 122)	11.9	2.8	12.0	7.0	3.7
MARI 35-5976 (Z-54, Fig. 123)	6.5	2.3	10.0	4.8	3.5

(*continued*)

Table 4.4.—continued

	Length, cm	Diam, cm	Row no.	Kernel characters	
				Width, mm	Thickness, mm
Race					
Chapalote (Mexico)	11.0	2.9	12.3	6.7	4.1
Specimen					
UM NA2141 (Z-16, Fig. 95)	9.7	3.0	8.0	6.8	4.8
MNA 6-6885 (Z-45, Fig. 115)	10.0	2.9	12.0	7.5	4.6
MARI 35-6004 (Z-55, Fig. 124)	16.4	4.0	16.0	7.5	4.4
Race					
Confite Puntiagudo (Peru)[c]	8.9	3.2	15.1	5.6	4.3
Canguil (Ecuador)[d]	11.1	3.3	15.0	5.6	4.5
Specimen					
AMNH 30/6334 (Z-36, Fig. 108)	—	3.4	16.0	5.0	3.9
Senter U5/4554	—	3.2	14.0	6.0	5.0
Senter U5/4552 (Z-56, Fig. 125)	—	3.2	14.0	5.5	5.2
Senter U5/4239	—	3.0	12.0	5.8	5.0
Race					
Mochero (Peru)[c]	7.6	4.1	14.0	7.6	4.5
Specimen					
AMNH 30.3/2301A (Z-24, Fig. 98)	7.5	3.6	12.0	7.5	4.0

Race					
Oke (Brazil)[e]	8.2	3.4	10.7	7.9	4.3
Specimen					
MNA 6894 (Z-46, Fig. 116)	—	3.5	8.0	7.8	4.3
Race					
Karapampa (Bolivia)[f]	12.2	3.0	8.8	7.9	3.3
Specimen					
MNA 6-6223 (Z-1, Figs. 82, 83)	—	3.1	8.0	7.5	3.6
AMNH 30.0/3 (Z-26, Fig. 100)	—	2.8	8.0	8.0	3.8
MNA 6-6755 (Z-44, Fig. 114)	—	2.0	10.0	6.0	4.2
Race					
Niñuelo (Bolivia)[f]	12.1	3.2	8.2	9.5	4.3
Specimen					
Paris 24.13.3645 (Z-41, Fig. 111)	6.2	3.3	10.0	9.3	4.3
Race					
Puya (Venezuela)[g]	17.1	3.8	13.6	7.1	3.4
Specimen					
Frissell 4889 (Z-43, Fig. 113)	—	3.9	12.0	6.0	3.6

a. Wellhausen et al. 1952.
b. Roberts et al. 1957.
c. Grobman et al. 1961.
d. Timothy et al. 1961.
e. Brieger et al. 1958.
f. Ramírez et al. 1960.
g. Grant et al. 1963.

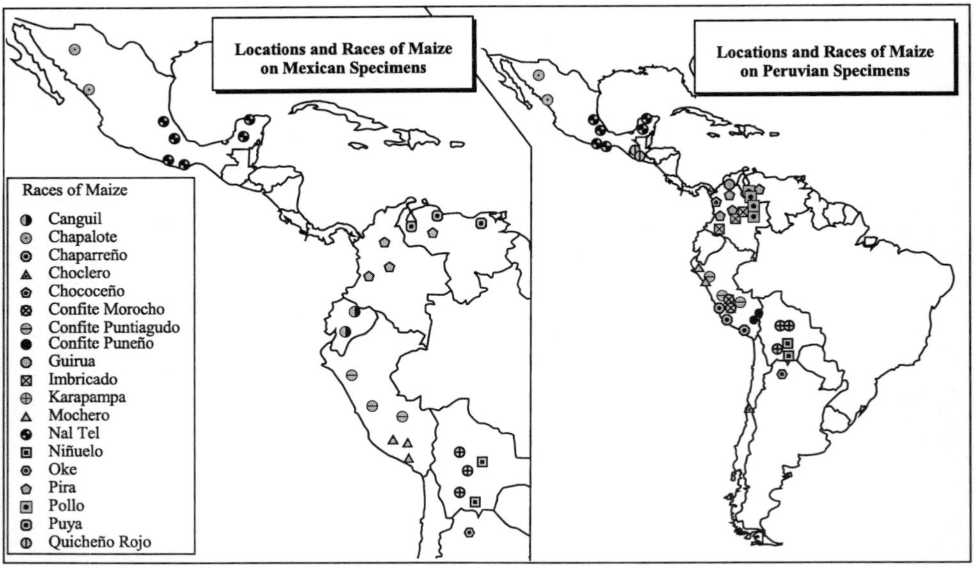

Fig. 126. Map showing geographical distributions of extant maize races identified on pre-Columbian pottery.

well as on the pottery replicas corroborates their identifications in the archaeological record. Although validity of the method of identifying archaeological maize by comparison to extant races has been questioned (Benz 1994), the results of this study show that prehistoric material from the Classic period is surprisingly comparable to extant maize races in most cases when using measurements from a complete ear with kernels intact. Each race is described below, beginning with those most frequently depicted.

Pira/Nal Tel. The maize race on most of the jars, twenty-nine from Peru (table 4.3) and thirty-six from Oaxaca (table 4.4), is classified as Pira/Nal Tel because both races are virtually indistinguishable and are the first and second choices for specimens in this group. Nal Tel is an ancient indigenous Mexican popcorn with small round yellow kernels and long glumes (Wellhausen et al. 1952). Pira is a primitive Colombian popcorn with small round white kernels (Roberts et al. 1957). The ears of both races are short, slender, and taper slightly from the base to the tip. Both mature very early after a couple of good rains. Although Nal Tel, still grown in the Yucatán Peninsula, is best adapted to low elevations, it grows at elevations up to 1,800 meters. Varieties similar to Nal Tel have been collected on the Pacific coastal plain in Oaxaca and Guerrero. Because it is early maturing, Nal Tel can be reliably grown in times of drought.

Pira is adapted to intermediate elevations from 400 to 2,000 meters and is found primarily in the Department of Cundinamarca, Colombia. It is believed to have once been grown more widely because it has been collected in distant regions ranging from Bolivia to Venezuela. L. M. Roberts et al. (1957, 40) suggest that Pira may be related to the Peruvian races Confite Morocho and Confite Puntiagudo, but they note that the ear shape and regular arrangement of the rounded white kernels are markedly different from those of the Peruvian popcorns. Nal Tel, Pira, and Confite Morocho were tightly clustered among the choices for specimens in this group. Ears with rounded kernels were identified as Pira/Nal Tel and ears with pointed kernels as Confite Morocho. The consistent statistical linkage of the three races supports the idea of Roberts et al. that Pira and Confite Morocho are closely related.

Pira's present distribution in Colombia is in the Department of Cundinamarca, growing at an average elevation of 1,100 meters. In Venezuela, its distribution is restricted to the state of Táchira, which borders Colombia (Grant et al. 1963, 36). It is geographically and genetically isolated and is centered in the Magdalena Valley at low altitudes. The maize on these jars is a primitive small round seeded popcorn. The Moche may have acquired Pira/Nal Tel through trade with neighbors to the north. It would have been well adapted to growing conditions on the dry North Coast of Peru. Pira/Nal Tel is seen in association with Ai-apaec, the Moche creator god, on twenty-five jars; in association with an unidentified human figure; on a jar depicting a mound of maize; and on six jars depicting a basket of maize. Representations of Pira/Nal Tel on Moche jars can be seen in figures 12, 15, 16, 17, 18, 23, 29, 30, 31, 32, 33, 40, 41, 42, 45, 46, 47, 50, 51, 52, 53, 54, 61, 63, 70, 77, 78, 79, and 80.

Pira/Nal Tel is the race most frequently depicted on Zapotec urns. The repeated representation of Pira/Nal Tel on Monte Albán IIIB and IV urns implies that this ancient primitive popcorn had special religious significance for the Zapotecs. Whether Pira/Nal Tel was the main type of maize grown for subsistence is open to question. The consistent depiction of Pira/Nal Tel on urns, along with its faithful representation in the actual botanical form, when all other aspects of the urns are highly conventionalized, strongly suggests that this race had important symbolic connotations. In contrast to the Moche, who realistically depicted many food plants on their pottery jars, maize was the only plant realistically depicted on Zapotec urns.

Nal Tel is an ancient indigenous race that matures after several good rains, making it highly valuable in a dry area. It may have been retained by the Zapotec for votive offerings because it is an ancient variety, descended from primitive maize such as that found in the Valley of Tehuacán (Mangelsdorf, MacNeish, and Galinat 1967a, 188). Although Nal Tel is not as productive as other maize races, it is dependable because it is better adapted to drought than many other races.

The name Nal Tel is from the name commonly used to refer to this variety of maize in the Yucatán Peninsula. Wellhausen et al. report that the name is of Mayan origin; *nal* means ear of maize, and *tel* means rooster. They indicate that since the rooster is an early riser, the name refers to the early maturity of this type of maize (1952, 62). (I realize the rooster is of Old World origin.) The Mayan origin of the words is confirmed by the *Diccionario de Motul,* but the meaning of the words is slightly different from the above interpretation. *Nal* refers to maize in the fresh state, that is, before it is dried (Martinez Hernández 1929, 662). *Nal* is what is commonly referred to as corn on the cob, and in Mexico it is sold in the markets boiled in the husks (Victoria Bricker personal communication). This interpretation also indicates that the name is used to designate an early maturing variety.

Nal Tel grows at elevations from 100 meters to 1,800 meters but is best adapted to lower elevations. Most modern collections of this race have been in the state of Yucatán and in neighboring parts of Guatemala. Ears virtually identical to Nal Tel from Yucatán were collected in the Huasteca area of the Gulf Coast, and other collections have been made in Oaxaca and Guerrero (Wellhausen et al. 1952, 60). Pira/Nal Tel is seen in association with the God of Glyph "L," Cocijo, the God with the Ear of Maize in His Headdress, the Opossum God, an unidentified figure, and urns encircled with ears of maize. Representations of Pira/Nal Tel on Zapotec urns can be seen in figures 3, 84, 85, 86, 87, 88, 90, 91, 92, 93, 94, 96, 97, 99, 102, 103, 104, 106, 107, 109, 112, 118, 119, 122, and 123.

Archaeological remains of Nal Tel have apparently been found in Oaxaca. Three cob fragments found at the Fábrica San José site in the Etla Valley were described as "Nal Tel-like" and reported to have been used for fuel (Ford 1976, 264). Nal Tel has been reported at the Preclassic Cuello site in Belize, where it appears as early as 2000 B.C. (Miksicek et al. 1981).

Confite Morocho. Confite Morocho was identified on thirteen Peruvian jars. It is an ancient primitive popcorn that may be a living counterpart of the earliest domesticated maize in Peru (Grobman et al. 1961). It has imbricated red or brown kernels, long glumes, and staminate spikelets in the upper part of the ear, a characteristic of the oldest archaeological maize found in the Valley of Tehuacán, Mexico, estimated to be at least five thousand years old. Confite Morocho is still found growing in the Andean highlands. Its center of distribution is in the Department of Ayacucho. The small ears on Moche jars may represent the earliest domesticated maize in Peru, and they resemble archaeological finds of such tiny maize cobs in association with religious or ceremonial objects that have been made at sites in the Department of Lambayeque on the far North Coast of Peru. At the same sites, cobs of larger, more productive races are found in domestic contexts, which suggests that the primitive small-eared race was grown for ritual purposes (Oscar Hernández Córdova personal communication).

Confite Morocho is found today at elevations of 2,500–3,000 meters above sea level. Grobman et al. (1961) postulated that it was once more widely distributed in Bolivia and Ecuador, growing at intermediate elevations and in the river valleys along the coast before it evolved environmental specificity. Because of its early maturity and adaptability to extreme environmental conditions, it could be grown in these areas. Alternatively, it may have been an exchange commodity in trading between the peoples of the highlands and the coast.

The Indians of the Andean highlands still grow Confite Morocho in small plots. It is used for consumption in its popped form. Because it flowers at a different time than other races, it continues to grow in relatively pure form. The Indians select ears for their small flinty kernels, which are particularly good for popping (Grobman et al. 1961). Confite Morocho appears in association with Ai-apaec on eleven jars and on two vessels depicting baskets of maize. Figures 10, 11, 13, 19, 27, 28, 36, 37, 38, 49, 55, 56, 72, and 73 illustrate vessels with depictions of Confite Morocho.

Oke. Oke (Brieger et al. 1958) is a rare race from a high valley in the Andes of northern Argentina referred to as the Quebrada de Humahuaca. This region was the southern boundary of the Inca Empire in pre-Columbian times. The Quebrada, situated above the main cloud zone, receives little rainfall but is well insulated by the cloud cover. Maize and potatoes are grown on terraced, irrigated plots. Maize adapted to these high elevations always has small, often egg-shaped ears with irregular rows. Although the Humahuaca area is only about seventy by thirty kilometers, there is unusually rich diversity of maize here. In small adjoining plots, farmers cultivate five or six different types of maize that are used for different purposes. Not much is known about Oke except that it is a black corn with rounded kernels that sometimes grade to slightly pointed. Its origin is obscure, and it is the only type of black corn found in Humahuaca. Oke is depicted on three Moche jars (P-7, fig. 14; P-12, figs. 21 and 22; P-14, fig. 24) in association with the Creator God Ai-apaec. It also occurs on a Zapotec urn (Z-46, fig. 116) in association with the Goddess "13 Serpent."

Chapalote. Like Nal Tel, Chapalote is an ancient indigenous race of primitive popcorn. Ears of Chapalote are distinct from Nal Tel in their pronounced cigar shape and brown pericarp (Wellhausen et al. 1952, 54–58). In the early stages of evolution, as evidenced by materials from the Valley of Tehuacán, these two races could be distinguished only by the colors of their pericarps (Mangelsdorf 1974, 117). Chapalote is found today in a restricted area in the states of Sinaloa and Sonora, but it is believed to have been more widespread in western Mexico and parts of the southwestern United States in prehistoric times. Chapalote that had been crossed with teosinte, a wild relative of maize, has been reported from caves in the vicinity of Mitla, Oaxaca, dating to Monte Albán IV (A.D. 620 to 740 A.D.) (Flannery and Smith, 1983, 206). Present collections of Chapalote are found at elevations from 100 to 600 meters, but it is adapted to a wider elevation range and will produce ears at an altitude up to 2,200 meters.

Chapalote is on four Moche jars: a stirrup-spout jar depicting Ai-apaec wearing a Moche warrior's helmet (P-36, fig. 48); an olla (P-48, figs. 59 and 60) depicting a human figure; a stirrup-spout jar (P-57, fig. 67) depicting a maize plant with a bird in the tassel; and a stirrup-spout jar (P-58, fig. 68) depicting a mound of maize. The bird is a *lorito* (parrot) (Grobman et al. 1961). *Loritos,* one of the worst pests of corn, are frequently depicted in association with maize in the pre-Ccolumbian art of Peru. Chapalote is also seen on three Mexican pieces: a Zapotec urn (Z-16, fig. 94) depicting the God of Glyph "L," an urn (Z-45, fig. 115) depicting the God with the Bowknot in His Headdress, and a sherd of two mazorcas (Z-55, fig. 124).

Pollo. Pollo was identified on two Moche jars. Pollo is a primitive popcorn grown in the Andean highlands. Collections of Pollo have been made in Colombia (Roberts et al. 1957) and in Venezuela (Grant et al. 1963). It has been identified among archaeological plant remains from the El Tiestal site in Venezuela (Mangelsdorf and Sonója 1965). Ears of the ancient indigenous Mexican races Nal Tel and Chapalote introgressed by Pollo have been identified from finds at Sitio Pití, an archaeological site in the Chiriquí highlands of Panama (Linares, Sheets, and Rosenthal 1975).

Today, Pollo grows relatively isolated in Colombia. Its center of distribution is Boyacá, the region where the archaeological remains of the Chibcha culture have been found. Roberts et al. (1957) hypothesize that Pollo was grown there by the cave dwellers who preceded the Chibcha. Ulysses Grant et al. (1963) emphasize that Pollo has striking similarities to the highland Guatemalan races Nal Tel, Tierra Alta, Serrano, and Quicheño Precoz. They suggest that Pollo was a result of cultural contact between the highland Maya of Central America and the Chibcha of the Andean highlands.

The identifications of Pollo on Moche and Zapotec vessels provide evidence that a race which is a close match for these two races had widespread distribution during the Classic period and contribute to the mounting botanical evidence for cultural contact throughout a wide area among different Latin American groups during that period. Pollo is on two Peruvian jars: one (P-24, figs. 34 and 35) depicting Ai-apaec with two companion figures and one (P-66, fig. 74) depicting a basket of maize.

Confite Puntiagudo/Canguil. Confite Puntiagudo and Canguil are two races whose identifications are interchangeable. Confite Puntiagudo, as its name signifies (Confite means popcorn and Puntiagudo means pointed), is a primitive imbricated popcorn believed to be derived from Confite Morocho (Grobman et al. 1961) with an admixture of *Tripsacum*, a wild grass related to maize. It has higher row numbers and longer ears than Confite Morocho. Kernels vary from strong to reduced imbrication, and on some ears the kernel rows spiral, a feature that is more typical of prehistoric ears from the coast of Peru. The Indians in Cuzco refer to Confite Puntiagudo as *Chili-Sara*, which is Quechua for "Old Corn." It is early maturing and one of the most widely distributed Andean races. It is found throughout the highlands of Peru and extends into the Andean regions of Colombia and Ecuador, where it has been described under the name of Imbricado (Roberts et al. 1957), and into Bolivia, where is it called Pisankalla, Pisingallo, or Pisincho (Ramírez et al. 1960). It is adaptable to a wide range of environmental conditions from elevations of 3,500 feet to sea level.

Canguil is a pointed white popcorn that grows in the mountains of Ecuador. It has been collected in the Andean highlands at a mean altitude of 2,260 meters (Timothy et al. 1963, 24) and resembles Pisankalla, a highland Bolivian race.

Representations of Confite Puntiagudo/Canguil appear on a Zapotec urn (Z-36, fig. 108) depicting Cocijo and on mazorca sherds (Z-56, fig. 125) excavated from an archaeological site near Xochicalco, Morelos (Donovan C. Senter personal communication, November 26, 1975). The sherds date to the Early Postclassic period, ca. A.D. 900–1100, and are remarkable for the shape of their kernels, which are strongly imbricated and distinctly beaked, features characteristic of Confite Puntiagudo/Canguil.

Representations of Confite Puntiagudo/Canguil on pottery models from Mexico signal that either there were cultural connections between Mexico and the Andes or this race was once grown in Mexico, where it is extinct today.

Karapampa. Karapampa was identified on three Zapotec urns (Z-1, figs.

82 and 83; Z-26, fig. 100; Z-44, fig. 114), one depicting the God of Glyph "L" and the other representing the God with the Ear of Maize in His Headdress. It was also found on four Peruvian jars: one (P-47, fig. 58) depicting a human figure and three (P-67, fig. 75; P-68, fig. 76; P-70, fig. 78) representing baskets of maize. Karapampa is an eight-rowed pop-flint variety of maize reported from the highlands of Bolivia (Ramírez et al. 1960). This race has a slender cylindrical ear with red or orange kernels that may be slightly imbricated. It is adapted to growing at elevations from 1,800 to 2,160 meters. It appears to have been transported to Mexico via an extensive trade network between Central and South America.

Chaparreño. An anciently derived coastal flour corn, Chaparreño (Grobman et al. 1961), has a long history in Peru. It is found at low elevations in the valleys between the Tambo River in the Department of Arequipa and the Rimac River near Lima. The soft kernels are yellow or white, occasionally purple. Although the kernels are not imbricated, they may be slightly beaked and deeply dented. The name derives from the valley of Chaparra in the Department of Arequipa, where the race is intensively grown. Chaparreño is depicted on two Moche jars, one (P-27, fig. 39) showing the Creator God Aiapaec with two companion figures, the other an olla (P-56, fig. 66) representing a frog. Chaparreño has been found at Nazca and Ica archaeological sites. The recovery of Chaparreño at La Rinconada near La Molina, Lima, confirms that it was already in existence by A.D. 1000 to 1200. Chaparreño and Mochero are closely related. Both are early maturing races well adapted to the dry desert conditions of coastal Peru.

Choclero. Choclero is a lowland flour corn from Chile (Timothy et al. 1963). The kernels are yellow, and they are somewhat pointed and beaked in some ears. Choclero refers to corn used for "choclos," or green roasting ears. It is hypothesized that Choclero derived from hybridization between a North American dent corn and a conical-shaped South American corn. The depiction of Choclero on three Moche jars (P-53, fig. 64; P-54, fig. 65; and P-55) indicates that it originated in prehistory by circa A.D. 500. All three are stirrup-spout jars depicting a rodent nibbling an ear of maize.

Mochero. Mochero is an early maturing race grown on the North Coast of Peru. It has a limited distribution today in the Departments of La Libertad, Lambayeque, and Piura. It is not grown at altitudes over 50 meters above sea level. Mochero has short, stubby ears which are cylindrical. They are tapering at the tip and rounded at the base. The kernels have up to fourteen rows and tend to be irregularly arranged so that the kernel rows appear in a spiral pattern. The name Mochero has been assigned to this race by Grobman et al. (1961, 201) because some of the first collections made were from the Moche Valley. Another name given this type of maize by farmers is Blando Local. On the North Coast, Mochero is prized for making chicha, corn beer (Nicholson 1960).

Mochero is depicted on a Moche stirrup-spout jar (P-61, fig. 69) representing a mound of maize and on a Zapotec urn representing the God of Glyph "L" (Z-24, fig. 98). Grobman et al. report that Mochero's "largest concentration is found in the small dry valleys of the 'Pueblos' area north of the city of Chiclayo, where it is irrigated by the summer or avenida waters of small, irregularly flowing rivers" (1961, 198). Chiclayo is on Peru's dry desert coast in the Department of Lambayeque. This drought-resistant race can be harvested three and a half months after planting and does not require irrigation.

Imbricado. Imbricado is a Colombian popcorn with distinctive pointed, overlapping kernels that is also found in Ecuador and Peru (Roberts et al. 1957). Three ears of Imbricado, each formed from the same mold, occur on a Moche piece (P-63, fig. 71) that represents a bowl of maize. This piece is not an actual container as are the other jars in this corpus. The maize ears have twisted rows and long, imbricated, overlapping kernels that are characteristic of Imbricado. This race has affinities with Guatemalan Imbricado (Wellhausen et al. 1957) and with Palomero Toluqueño from Mexico (Wellhausen et al. 1952). Mangelsdorf (1974) considered Palomero Toluqueño to be the ancestor of all pointed South American popcorns.

In addition to the vessels depicting maize, a two-part mold is illustrated in figure 2. The provenience of this mold is the Valley of Chancay on the Central Coast of Peru. This piece is important because it clearly documents the use of molds for anyone who might otherwise question this basic premise of the research. On the basis of the negative imprint of the ear, this mold most closely resembles the highland Guatemalan race Imbricado, which has twelve to sixteen rows of pointed imbricated kernels on long tapering ears (Wellhausen et al. 1957). The mold has twelve rows of imbricated kernels, and the ear measurements are virtually a perfect match for Guatemalan Imbricado (Dunn 1977, 103).

Whereas most of the depictions of maize are of only half the ear, this mold was meant for fashioning a complete ear. A section of the mold has a hole in the center. This mold could have been used to form a vessel in the shape of a complete ear with the hole providing for the attachment of the spout to the vessel body. A vessel of this description is illustrated by Grobman et al. (1961, fig. 130).

Güirua. Güirua, a Colombian race (Roberts et al. 1957), is restricted to the Department of Magdalena near the Sierra Nevada de Santa Marta, an area formerly inhabited by the Tairona culture, which is believed to have received strong influences from Central America and Mexico (Loven 1935; Reichel-Dolmatoff 1965). The origin of Güirua is unknown. The meaning of the name is unknown, but the name is presumed to be of Indian origin because it was given to this race by the non-Spanish-speaking Indian farmers from whom it was collected (Roberts et al. 1957).

Roberts et al. (1957) assign Güirua to the category "Races Probably Introduced," which means that it has no obvious progenitor in Colombia but has a counterpart in another country. It is interesting that Güirua's counterpart is Negro de Chimaltenango, a highland race from Guatemala (Wellhausen et al. 1957). Whether this race diffused from Central America into Mesoamerica, as Mangelsdorf proposed (1974), cannot be determined on the basis of present evidence, but the presence of this race in Colombia and highland Guatemala strongly points to cultural contact between the two areas. Güirua's representation on a Moche jar (P-60) suggests that such connections extended as far south as Peru during the Classic period.

Puya. Puya is a Venezuelan race (Grant et al. 1963) depicted on a Zapotec urn representing the God with the Ear of Maize in His Headdress (Z-43, fig. 113). Puya is a dent corn with hard yellow kernels that has a counterpart in Colombia. Its distribution is centered around Lake Maracaibo and extends along the Caribbean coast and up the Magdalena River, a region inhabited by Arawak tribes. The Arawak Indians introduced Puya into the West Indies before the Spanish arrival in America. Since there are no antecedent yellow dent races in South America but they are prevalent in Mexico, the direction of movement was probably from Central America into South America. Puya is believed to be an ancient introduction from the Gulf of Mexico into Colombia and Venezuela, where it was modified by introgression with South American maize.

Niñuelo. Niñuelo (Ramírez et al. 1960) is an early maturing Bolivian race adapted to elevations from 1,800 to 2,000 meters, with round flint, white kernels. It is depicted on a Chimú olla (P-51, fig. 62) representing a human figure and on a Zapotec urn depicting the God with the Ear of Maize in His Headdress (Z-41, fig. 111).

Confite Puneño/Quicheño Rojo. Confite Puneño (Grobman et al. 1961), a highland Andean race with short ovoid ears, is the smallest plant of all the Peruvian races. It is grown at 3,600–3,900 meters above sea level, the highest altitudes of any maize in the world. Its distribution is the Collao plateau around Lake Titicaca. Except for being slightly larger in size, Confite Puneño is virtually identical to Confite Iqueño, an archaeological race found on the Peruvian coast that dates to twenty-five hundred years ago.

Quicheño (Wellhausen et al. 1957) is a medium to late maturing flint popcorn with round kernels from Guatemala. It grows at elevations ranging from 1,676 to 2,438 meters above sea level in Guatemala. Sub-race Rojo, collected at Quetzaltenango and Huehuetenango in El Quiché at 1,830 meters, has red kernels and is believed to represent a relict primitive form of this race. The small, slightly tapering ears with rounded butts, slender cobs, and deep red isodiametric kernels closely resemble prehistoric ears excavated from archaeological sites in Peru. Quicheño has affinities to Confite Morocho and is believed to be a prehistoric introduction from South America into Guatemala.

Confite Puneño/Quicheño appears on five Moche jars (P-11, fig. 20; P-12, figs. 21 and 22; P-15, fig. 25; P-16, fig. 26; P-42, fig. 53) in association with Ai-apaec and on a Moche stirrup-spout jar (P-59) depicting a mound of maize.

Chococeño. Chococeño is found growing today on the Pacific coast of Colombia and Ecuador (Timothy et al. 1963). A form of Chococeño introgressed with Coastal Tropical Flint is also grown in the Darien region of Panama by a group of Chocó Indians who migrated from Colombia more than twenty years ago (Nickerson and Covich 1967).

Chococeño is an unusual race because of its extremely tripsacoid characteristics and the way it is cultivated. It has thick short conical ears with a high row number of narrow thin kernels. It is grown by a primitive method of cultivation in Ecuador and Colombia in a region of very high rainfall, sometimes exceeding 400 inches in one year (Roberts et al. 1957). In the area for planting, the small trees and bushes are cut over, then the seed is sown out over this area. It germinates on top of the ground with the plants growing up through the cut-over vegetation. The ears encircling a stirrup-spout jar representing Ai-apaec (P-32; figs. 43 and 44) resemble Chococeño.

Dent corn. The vessel in figure 121 is encircled by ears of maize whose kernels show a slight indentation, that is, denting. Denting is a characteristic of Tuxpeño, a race considered ancestral to the highly productive dent corn grown in the Corn Belt of the United States (Mangelsdorf 1974, 191). The ears on this vessel closely resemble Tuxpeño, as suggested by Wellhausen et al. (1952, 148). The vessel provides evidence for the existence of dent corn in Oaxaca as early as ca. A.D. 600. Other evidence for the presence of prehistoric dent corn has been shown by finds in association with the Fremont culture in Utah dating from A.D. 900–1200 (Galinat and Gunnerson 1963, 154).

In summary, there is diversity in the types of maize, the range of climates and environments to which they are adapted, and the countries in which the modern collections have been made. The maize represented on the pottery suggests that some races were more widespread in prehistoric times. If present distributions are indicators of the extent of prehistoric distributions, the findings signal that during the Classic period the North Coast of Peru was a center for contact and exchange between diverse cultural groups from a wide geographical area.

5

Prehistoric Races of Maize and Cultural Diffusion

A total of nineteen races of maize were identified on 129 specimens (refer to tables 4.1, 4.3, and 4.4). Sixteen races are represented on seventy-two Peruvian jars, and ten appear on fifty-six Mexican urns and mazorca sherds. Eight of the same races appear on jars from both regions. These races represent fully the extensive biodiversity of maize. There are pop, flint, flour, and dent corns. There is a rainbow of colors including white, yellow, red, purple, brown, blue, and black. There are ears with few kernel row numbers ranging from eight to twelve and others with as many as twenty-two rows of kernels. Extant forms are found today in Mexico, Guatemala, Venezuela, Colombia, Ecuador, Peru, Bolivia, Argentina, and Chile (see fig. 127). They are adapted to a range of habitats and environments from the seacoast to the highest elevations in the Andes Mountains where it is possible to grow corn. The phenology includes the earliest maturing races that flower less than 70 days after planting and late maturing types that require more than 140 days to flower.

The Latin American maize races have been classified into fourteen major groups (Goodman and Bird 1977). Seventeen races identified on the terracotta models can be assigned to six of those taxonomic categories: the Northern South American Popcorns, the Lowland Flours, the Chapalote Group, the Northwestern South American Races, the Central Andean Complex, and the Caribbean Dents. Two races were not assigned to any of the fourteen taxonomic categories. Choclero from Chile was so distinct it did not fall in any of the groups (Goodman and Bird 1977, 207), and Quicheño from Guatemala was not included in Goodman and Bird's (1977) data set.

Six races fall into the category Northern South American Popcorns. This category contains three subgroups: yellow popcorns, white popcorns and small flints, and pointed popcorns. All three subgroups are represented on the pottery. Nal Tel and Confite Morocho are yellow popcorns; Pira is a round white popcorn; and Confite Puntiagudo, Canguil, and Imbricado are pointed popcorns. Goodman and Bird (1977, 210) commented that "the inclusion of Nal-Tel with this group seems especially doubtful." The fact that identifications of Pira, Nal Tel, and Confite Morocho were tightly linked in all examples of these races on pre-Columbian pottery, however, supports inclusion of Nal Tel in this group and suggests that Nal Tel, Pira, and Confite Morocho all represent an ancient primitive popcorn that was geographically widespread in prehistory, ranging from Mexico into northern South America.

Güirua from Colombia is the sole example representing the Lowland Flour Corns. It belongs to a group of highly colored flour corns grown around the periphery of the Amazon and Orinoco basins. Chapalote is the sole example representing the Chapalote group of small-kerneled flint/popcorns restricted to northwestern Mexico.

Three of the six extreme types that make up the Northwestern South American group are represented on the corn pots. First, there is Pollo, a small-eared, large-kerneled, flinty popcorn. The second type is Chochoceño, a popcorn that typically has short conical ears with high row numbers. Third, Chaparreño and Mochero fall into the subgroup North Andean floury maize.

In the South Andean flint category are Karapampa from Bolivia and Confite Puneño from Peru. There is one example of Niñuelo which was assigned to the Cuzco group. Goodman and Bird (1977) questioned this assignment, and since the Niñuelo identification on the pottery clusters with Araguito and Nal Tel from Central America, it is possible this race is closely related to the northern South American popcorns. Oke is a highland maize in the Central Andean Complex, a group of races with specialty uses for specific foods, drinks, or dyes. Puya is the sole example of a Caribbean dent corn which is thought to have arisen from hybridization between Mexican dent corns and northern South American flints. Two races identified on the pottery, Quicheño and Choclero, were not among the 219 races grouped by Goodman and Bird (1977).

Some of the races identified provide evidence for cultural contact between Mesoamerica and South America during the Classic period, ca. A.D. 300–900. Nal Tel and Pira, which are considered one and the same for the purposes of this study because they are indistinguishable morphometrically and phenetically, are the races most frequently found on Oaxacan and Peruvian vessels. Nal Tel is grown today primarily in Yucatán and Guatemala. Pira is found today in Colombia and Venezuela. Pollo, Güirua, and Imbricado, Colombian races identified on the pottery, have affinities with Mesoamerican races. Pollo was identified on three Zapotec urns and two Moche jars. This suggests that the Zapotec and Moche cultures were in contact, either directly or via an intermediate cultural group from which both were receiving trade goods.

The identifications of the South American races Confite Puntiagudo/Canguil, Oke, Karapampa, and Puya on Zapotec urns and early Postclassic mazorca sherds from Morelos, and of the Mexican race Chapalote on three Peruvian jars, suggest that these races had spread into Mexico and northern South America, respectively, as early as A.D. 900, implying prehistoric cultural contact and exchange between these regions.

In view of these findings, a comparative analysis of the pottery depicting mold-made maize and a review of other evidence for cultural contact between the two areas are considered and discussed to demonstrate corroborative evidence for contact and to elucidate the nature of diffusion. The findings are also discussed in terms of potential contributions to our understanding of the evolution of races of maize in the archaeological record.

Comparative Analysis of the Pottery Vessels

The pottery and vessel forms do not provide any evidence for direct contact. The Zapotec pottery is predominantly gray ware. The Moche pottery is predominantly orange ware with some painted decorations in red, white, and black. The pottery from both areas is generally well fired and highly polished. The Oaxacan urns are almost all cylindrical with the figure built out in front of the container. The Moche jars are the plain jar form, the "basket" olla, and the stirrup spout. Depictions of figures are incorporated into the body of the vessel.

There are four exceptions among the Oaxacan pieces in which the body of the figure is incorporated into the vessel body in the method of fashioning on Moche jars. These take on a more characteristic bottle shape illustrated in figures 92, 102, 103, 107, and 112 (see also Eubanks 1973, plates 21, 30, 36 and 44). The pottery of these rather unusual jars ranges from brown to dark gray or black. Two have proveniences of Zaachila; one is from the Mixteca and another from Ejutla. Since none are purportedly from Monte Albán, and they represent what may be considered provincial wares, the difference in execution is thought to represent regional variation in style rather than an import. Furthermore, the maize identified on these vessels is not unusual but is the race identified on most Oaxacan specimens, Nal Tel.

Other pottery from Oaxaca that may signal adaptation by the Zapotec of a foreign idea in the manufacture of pottery are vessels surrounded by maize ears (figs. 121, 122, 123, 124, and 125). Included are three vessels and a sherd of two maize ears. It is assumed that the latter was probably from this same type of vessel because of the shape of the interior wall behind the maize ears. These vessels are made of the same well-fired gray ware as most other Zapotec urns. The vessel shape, the method of displaying the full length of the maize ear rather than a section of it, and some of the maize races identified suggest that these pots reflect influences from outside Oaxaca. The circular shape with vertical walls may be related to tripod bowls and incensarios from the Guatemalan highlands (Rands and Smith 1965, 112). Similar types have also been found in burials on the Pacific coast of Guatemala (Shook 1965, 183).

Maize on Zapotec urns is almost always associated with a deity or ruler. Races of maize have been identified on urns associated with eight different figures from the Zapotec pantheon. On most of the Moche jars, maize is associated with a single divinity, Ai-apaec, the Moche Creator God. Similarities between these figures include solemn, staring, expressionless visages and representations of jewelry, including wrist bands, ear ornaments, and bead necklaces. One Moche jar (fig. 40) is unusual because the figure is apparently seated cross-legged. This posture is typical of the majority of the figures on Zapotec urns.

The most frequently depicted Zapotec figure symbolizes Cocijo, the God of Lightning and Rain. There are several Cocijo urns with maize in the headdress. The main attributes associated with Cocijo are the elements of the face composed of the bifurcated tongue protruding from the open jaws of a serpent and the nasal and buccal bands that represent feline characteristics. The most important attribute of Ai-apaec is the fangs, which supposedly denote the feline characteristics of the Moche god. A close association of this deity with a serpent is evident from the numerous representations of Ai-apaec wearing snake-head ear ornaments and the jar with a serpent curling around it (see figs. 43 and 44). Although fangs do not occur on any of the Zapotec urns with representations of maize, there are examples of figures with fangs among the categories of Cocijo, the God with the Buccal Mask of the Serpent, the Jaguar God, and the Bat God (Boos 1966a; Caso and Bernal 1952). Although the figures associated with mold-made maize do not provide any evidence for a direct relationship between the two cultures, they definitely share certain basic underlying religious concepts associating feline and serpent attributes with the most important divinities, and these, in turn, have an association with maize.

Miguel Covarrubias traces the origin of the various Mesoamerican rain gods, Cocijo, Chaac, and Tlaloc, back to the Olmec jaguar deity in the Preclassic period (1957, 60). Parallels between the Olmec feline and the feline representations in the Formative Chavín culture from Peru have been observed by numerous authors (Benson 1972b). It has been postulated that there are direct relationships between the Olmec and Chavín cultures and that Chavín derived from the Olmec (Spinden 1917; Strong 1943). The Moche fanged deity derived from the Chavín feline god (Rowe 1962b). I propose that there may be an indirect historical relationship between the Zapotec rain god and the Moche feline deity which can be traced to antecedent forms in the Olmec and Chavín cultures, respectively, but there does not appear to be any direct relationship between the Zapotec and Moche deities.

The technique of using the mold to give impressions of maize that are precise replicas of the natural ears is practically unique to the Moche and Zapotec Classic period cultures. A few isolated occurrences of representations of mold-made maize on the pottery of other cultures have been noted. These include an Aztec brazier (see fig. 120), a Mixtec urn, mazorca sherds from Morelos (see fig. 126), and some Chimú vessels. There is no evidence, however, that these representations were produced on a large scale, and these are all Postclassic period manufactures. Another example is a mazorca on an urn fragment from Lake Amatitlán in highland Guatemala (Nicholas Hellmuth personal communication, June 9, 1974) for which I have no information regarding chronological placement.

The question of whether this particular technique was an independent invention or whether it originated in one place and diffused to the other has been raised. Mangelsdorf observed: "This is a highly sophisticated form of ceramic art that so far as I know was not generally practiced with native plant products in any other part of the world" (1974, 188). Although cultures all over the world have depicted plants in vase paintings and stone carvings, incised or embossed on metals, woven in cloth, and in other artistic media, this molding technique appears to be virtually unique to these two cultures of the New World at this early time in history. The closest examples to the maize casts are impressions of wheat accidentally fired in Iron Age pottery from Europe and the Near East and used to identify the grains (Helbaek 1959, 368).

Considerations bearing on the problem of independent invention or a single origin include whether there are local antecedents in the receiving culture, contemporaneity of the two items, possibility for physical contact, and the number and complexity of the artifacts shared. Contemporaneity of the use of the mold-made maize replicas has already been established.

The depictions of fruits and vegetables on the pottery of the North Coast of Peru appears in the Early Formative among Cupisnique grave goods ca. 1000 B.C. (Larco Hoyle 1966, 51). Molded forms appear in the Vicús ceramics by the Middle Formative, ca. 600 B.C. (Larco Hoyle 1966, 223). The use of molds to depict fruits and vegetables in their form true to nature has a long antecedent development in Peru. Furthermore, the technique is used to represent at least eleven different plants. In Oaxaca, the only plant depicted by this method is maize. The technique was adopted suddenly during Monte Albán IIIB, and there is no demonstrable evidence that it was used by earlier cultures in Mesoamerica. Its abrupt occurrence in Oaxaca is apparently without any local antecedent development.

Elsewhere in Mesoamerica, the use of the mold first appeared at Teotihuacán during the Classic period, ca. A.D. 450. At this time molds were used to stamp out figurines and ornaments that were applied to pottery vessels (Cook de Leonard 1971, 184). There is no earlier use of molds in Mesoamerica. Molds for making figurines are the distinguishing marker between Teotihuacán II and Teotihuacán III (Covarrubias 1957, 136). During the late Classic period the making of mold-made figurines spread throughout Mesoamerica. It occurs in the Olmec area of the Gulf Coast, in the Huasteca region, and in Maya pottery of the Tepeu phase at Uaxactún (Willey 1947, 85). In contrast to this sudden appearance of mold-made ceramic objects in Teotihuacán III in the Valley of Mexico and its subsequent rapid widespread dissemination in Mesoamerica, mold making began on the North Coast of Peru more than a thousand years earlier and gradually gained popularity there as a method of pottery manufacture over hundreds of years.

Because the races of maize identified and the sudden appearance of molds in Mexico during the Classic period suggest contact between Mesoamerica and South America, it is necessary to review the possibilities for physical contact between the two areas and to look at other artifacts that may represent diffused traits or traded objects.

Other Evidence for Culture Contact

The evidence for trade between Mesoamerica and South America and proposed routes for interchange between these two areas has been thoroughly reviewed by Chester Chard (1950).

Possible means for pre-Columbian travel were overland by foot, by sea in canoes or sailing rafts, or a combination of both. There are three possible routes from Mesoamerica to South America: by canoe along the Caribbean coast, overland, and by sea along the Pacific coast. By sea along the Caribbean would have been relatively easy in dugout canoes traveling close to shore. The winds and currents do not present any major obstacles to travel along this route, and Chard considers this an established trade route from the Yucatán Peninsula along the Caribbean coast of Colombia and Venezuela (1950, 2). An overland route would have been considerably longer and more difficult because it extends through mountains and jungles, but it was possible. Chard also believes the Caribbean sea route was an established one from the Valley of Mexico to Panama.

Maritime trade along the Pacific coast would have been considerably more difficult because of strong currents, a rugged shoreline, and rough seas. Chard considers seafaring trade along the west coast of Mexico and Central America south to the coast of Ecuador as conjectural, although he concedes the possibility of limited, sporadic seacoast trade from Panama south to the coast of Ecuador (1950, 2).

Clinton Edwards, in contrast, considers maritime trade between the western Guatemala coast south to the Manabí coast of Ecuador and perhaps as far south as the Chiclayo Valley on the North Coast of Peru via large dugout canoes or sailing rafts a distinct possibility (1969, 4). He maintains that the actual conditions of the currents and the winds are not always as severe as have been reported and that at certain times of the year, lengthy sailing voyages on the open sea would have been feasible.

It has been demonstrated that conditions along the Pacific coast are not always as difficult as Chard has described, and at certain times of the year, travel by sailing crafts would be entirely possible. From January to March, there are northerly winds that make rapid sail from as far south as the Gulf of Guayaquil to ports farther north in Central America relatively easy. The return trip is possible, but it takes two or three times as long (Murphy 1941, 9). In his study of different means of aboriginal navigation, Edwards has convincingly shown that there was prehistoric sea travel along the Pacific coast by means of large dugout canoes and sailing rafts of lightweight balsa wood (1965a). Furthermore, the large balsa log sailing rafts of Guayaquil, which were in use on the coasts of Colombia, Ecuador, and northern Peru, were capable of making long sea voyages (Edwards 1960, 1965b).

Evidence from ethnohistorical accounts supports the contention that there was seafaring trade along the northern coast of South America at the time of the arrival of the Spanish. The first description of an encounter of an Indian sailing craft in the Pacific was written by Francisco de Xeréz. It described an encounter by Bartolomeo Ruíz, a captain under the command of

Francisco Pizarro, of a large raft with a sail carrying twenty men, loaded with cloth and other goods for trade off the coast of Ecuador (Porras Berrenechea 1937, 66). Other historical evidence bearing directly on the possibility of sea trade between Mesoamerica and South America is a translated Spanish document that contains a letter from the royal accountant Rodrigo de Albornoz to Charles V, dated December 15, 1525. This letter describes reports by Indians on the Zacatula coast at the mouth of the Balsas River on the west coast of Mexico of islands in the south that are rich in pearls and precious stones. Indians from the south were said to have traveled northward in large canoes bringing luxury items to trade for local products. The Indians from the south were said to have stayed for as long as five or six months to await good weather and calm seas before leaving to go back south (West 1961, 133).

Although it can be demonstrated that there was long-distance trade by the Aztecs, Maya, and Inca that extended into parts of Central America when the Spanish first arrived, answers to the questions of the antiquity and the nature of such distant contacts between the two areas must come totally from archaeology and depend on interpretations of remains of material culture. Inferences concerning sea travel by peoples along the Pacific coast of northern South America for long distances must be made with caution. The Moche only depicted small fishing boats; there are no illustrations of large sailing rafts laden with trade goods.

Evidence for trading contacts between Mesoamerica and South America has been considered by several authors, and it consists primarily of sporadic lists of cultural traits found in both areas. On the Formative time horizon, Michael Coe has compared pottery from La Victoria on the Pacific coast of Guatemala to Valdivia ceramics from the coast of Ecuador. On the basis of shared traits such as iridescent paint and pottery shapes, he has proposed that the two areas were directly connected through maritime trade (1960, 384). Also, on this early time horizon, Muriel Porter has compared features including the stirrup-spout, color zoning, zoned rocker stamping, flat and roller clay stamping, the feline motif, and dual representations from Tlatilco in the Valley of Mexico to traits from the Chavín culture of Peru. She suggests that there may have been connections between the two areas (1953, 372). Other traits believed to have derived from northwestern South America include tripod vessel bases, large hollow figurines, negative or resist decoration, and shaft and chamber burial tombs (Taylor 1970; Bell 1971).

The evidence for contacts during the Classic period includes trait lists and some references to specific traits. Alfred Kidder II observed similar traits such as trophy heads, human sacrifice, tripod vessels, metallurgy, negative painting, and a variety of pottery shapes (1940, 459). Stephen Borhegyi expanded the list to include sixty traits shared between different regions in Mesoamerica and South America. He observed Mesoamerican influences in archaeological materials from the Guatemalan highlands and the Pacific coast of Ecuador and postulated that contact by sea occurred between the two areas during the Classic period (1959).

Some specific traits that have received special attention are metallurgical techniques (Willey 1955, 41–42; Meighan 1974; Lechtman 1980), copper items from West Mexico (Mountjoy 1969), shaft tombs in West Mexico (Meighan 1969, 15), a thin orange vase in Ecuador (Jijón y Caamaño 1949), decoration of teeth by inlay and filing (Saville 1913), representations of the chimera in art from Peru and West Mexico (Meighan 1969, 18), nonfunctional copper axes from Oaxaca (Lothrop 1932, 258), emeralds from South America at Coclé in Panama (Lothrop 1937), and a Mixtec tripod vase with a painted Peruvian design (Covarrubias 1957, 196). Alana Cordy-Collins (1994) argues that the representations of hairless dogs that first appear in South America on Moche pottery is strong evidence for contact between West Mexico and the North Coast of Peru during the Classic period. These dogs are endemic to Mexico, and artists in the Colima culture in West Mexico made jars depicting hairless dogs as early as 250 B.C.. A West Mexican bird, the painted jay, is also postulated to have been transported from South America by pre-Columbian traders (Haemig 1979).

These examples show that there are many instances of items or traits found in Mesoamerica or South America that suggest contact between the two areas. Such objects and traits, however, are sporadic, and it may be argued that contacts reflected by the occurrence here and there of trade items do not bear directly on the problems of origins, history, and development of the cultures involved. There is, however, a growing body of evidence for contact between Oaxaca and the coasts of Ecuador and northern Peru. Several objects were found at Monte Albán that may be interpreted as Peruvian imports. One is a gold crown with a single gold feather which is identical to the gold crowns worn by Moche and Chimú dignitaries. This is the only example I know of one of these crowns found anywhere in Central America or Mexico (Lothrop 1936, 72–73). Along with this crown, small squares of differing sizes of sheet gold with a single small hole in the middle near the upper edge, identical to thin gold squares the Moche sewed on cloth to cover the robes of dignitaries, have been found at Monte Albán (Caso 1969, Lám. 24). Other imported items at Monte Albán include a copper object fashioned like a pair of large scallop shells with a Peruvian design (Lothrop 1936, 72–73) and thin nonfunctional copper axes that are commonly found on the North Coast of Peru (Lothrop 1936, 258).

The use of molds in pottery manufacture and metallurgy appeared sometime in the Late Classic period in Mesoamerica without precedent anywhere in Mexico or Central America. These appear to be technological innovations that moved northward from South America, where both have a long history of development. The surgical procedure of trephination may also have been derived from South America. Caso reported a trephined skull from Tomb 80 and two others from Tomb 84 in excavations at Monte Albán (1938, 38–39). These three, plus two others, were found in Monte Albán IIIB levels at that site. Skull surgery was practiced extensively in Peru, but the Monte Albán skulls represent the only examples of trephination in Mesoamerica, and they are found only in the Monte Albán IIIB period, the same time when the mold-made maize impressions suddenly appeared on Zapotec urns. Richard Wilkinson argues against the possibility that the Monte Albán trephined skulls are evidence of diffusion of the surgical technique from Peru. Since the techniques used in each place were different, he concludes that trephination at Monte Albán developed independently of its practice in Peru (1975). Because the custom does occur in both places, this may be an example of diffusion of the idea without the technique. Trephination in Mesoamerica was apparently restricted to Monte Albán during the Late Classic period, and at this same time other evidence for contact with South America seems to cluster at Monte Albán.

In review, the four criteria for culture contact are met. There are local antecedents in Peru for traits that appeared in Oaxaca during the Late Classic period, and there are no antecedents for the traits in Mexico or Central America. These traits include the use of the mold to display maize realistically, the presence of certain gold and copper items that do not appear anywhere else in Mesoamerica, and the practice of trephination which is without precedent outside of South America. All of these occurred at the same time period in the same place. Because these traits appear in Oaxaca but are not found in intervening areas, I propose that they are the result of direct contact by means of maritime travel along the coast from northern Peru to western Mexico. I have discussed the possibilities for sea travel and have demonstrated that such travel was at least probable in the Late Postclassic period before the arrival of the Spanish in the New World. That there was such travel five hundred years earlier cannot be demonstrated by the archaeological record. The botanical evidence complements the archaeological evidence. Five South American races of maize have been identified on Mexican specimens. These include Pollo from Colombia, Confite Puntiagudo from Peru or Canguil from Ecuador, Puya from Venezuela, Oke from northwest Argentina, and Karapampa from Bolivia. Canguil, Confite Puntiagudo, Oke, and Karapampa have no intermediate counterparts and are not found today in Mexico. The primitive Mexican popcorn Chapalote occurs on three Peruvian vessels. Confite Puneño identified on Moche jars has a counterpart from Guatemala, Quicheño Rojo, and this may signal indirect diffusion by way of highland Guatemala.

If the direction of diffusion can be inferred on the basis of the presence of local antecedents in a region, as proposed by Edward Lanning (1963), then it can be inferred that there were movements between South America into Mesoamerica. Presumably, if the contacts were a result of trade, goods and ideas would have been exchanged in both directions. The question of intercontinental contacts is a historical problem. It bears on cultural origins, dispersals, and the development of civilization. This study brings together data that indicate the existence of cultural contacts. Future studies and archaeological explorations should yield increasing evidence for a better understanding of the nature of such contact and its influence on cultural development during the Classic period.

Evolution of Maize: The Archaeological Record

Peru

Archaeological plant remains on the Peruvian coast have been well preserved because of the extremely dry desert climate, particularly on the south coast, where large collections of complete ears have been found. Archaeological finds of maize in the highlands are from Ayacucho (MacNeish et al. 1970). Therefore, knowledge of the evolution of maize in Peru is based primarily on the archaeological record from the coast.

The maize finds from Ayacucho are the oldest macroscopic remains found in Peru. The radiocarbon dates for the Chihua phase to which these remains have been assigned range from 4300 B.C. to 2800 B.C. (MacNeish et al. 1970, 38). It was originally believed that this maize was a primitive prototype of the Peruvian race Confite Morocho. Upon reassessment of the materials, however, Galinat proposed that the remains represent a separate, more primitive race which he calls "Ayacucho." Galinat postulates that Ayacucho is ancestral to many of the indigenous Peruvian races (1972, 107–8). No descriptions of the Ayacucho maize remains have been published. Therefore, it is impossible to compare it to other archaeological collections.

Other early preceramic maize has been reported from the Huarmey Valley on the North Coast at a site called Los Gavilanes (Bonavía 1982). Carbon 14 dating of plant, tassel, and cob material yielded a date of 3,750–100 years B.P. (Grobman and Bonavía 1978). Most of the cobs had eight rows, a few had ten, and one had twelve rows. The maize was a popcorn identified as an early, smaller form of Confite Morocho. Another find of preceramic maize was reported from Huarmey on the Central Coast (Kelley and Bonavía 1963), but no descriptions have been published. Maize has been recovered at several archaeological sites in Peru along the coast in the Moche, Casma, Chicama, and Supe Valleys and at a few highland sites in addition to Ayacucho in Peru. There are also relatively recent reports of archaeological maize from Chile and northwest Argentina. See Deborah Pearsall's work (1994) for a list and summary of these finds. They are not alluded to directly here because reports were either inaccessible, finds were too fragmentary or of charred materials, or information available was not sufficient for comparison to the maize ceramic models.

A collection of 732 cobs found at the Huaca Prieta site in the Chicama Valley on the North Coast of Peru has been studied by Robert Bird and Junius Bird (1980). These materials span the chronological periods Chavín-Cupisnique (ca. 1050 B.C. to 300 B.C.) and Gallinazo (ca. A.D. 20 to A.D. 220). Six types of Gallinazo maize were described: Type 1 has fourteen to fifteen kernel rows and resembles the archaeological race described by Grobman et al. (1961), Confite Iqueño; Type 2 with thirteen to fourteen rows resembles small Andean flint and flour corns; Type 3 with short narrow cobs and approximately ten rows resembles Confite Morocho; Type 4 has nine to ten rows with wider, slightly longer cupules; Type 5 resembles Type 2 but has cobs with the greatest length and width of the sample, and Type 6 has thirteen to fourteen rows with long deep cupules. Types 3 and 4 are most similar in size and shape to the oldest maize from the Valley of Tehuacán (Mangelsdorf et al. 1967a). Types 3 and 4 resemble pottery models identified as Pira/Nal Tel and Confite Morocho and should correspond to the Moche 1 category proposed by Bird (1984). The earliest Huaca Prieta maize from the Cupisnique period is distinguished by higher row numbers up to twenty (Bird 1978b). It is postulated that the Gallinazo maize with fewer rows and larger kernels was introduced to the Peruvian coast from the north or east (Bird and Bird 1980).

Margaret Towle (1952a, 1952b, 1954, 1961) described the evolutionary sequence of coastal races. There appear to have been basic differences between maize types from the South Coast and varieties from the North Coast. In general, South Coast maize has ovoid spherical ears with a high kernel row number. Most ears have sixteen or eighteen rows of isodiametrical kernels. There is a distinct evolutionary trend in South Coast maize from ca. 500 B.C. to A.D. 800 toward increased size of ear length and diameter. There is also a change in kernel arrangement from irregular spiraling rows to straight rows with a reduction of kernel row number.

Approximately contemporaneous North Coast maize collections contrast with trends exhibited by South Coast maize. The small ears are cylindrical and have characteristically low kernel row numbers. North Coast maize ears tend to have eight to ten rows. During the same period of more than a thousand years, North Coast maize exhibited little change morphologically and only a slight increase in size.

Grobman et al. (1961) proposed that the races Confite Iqueño, Confite Morocho, Kculli, Rabo de Zorro, Uchuquillo, Confite Puneño, Chullpi, Confite Puntiagudo, and Enano had appeared in Peru by the end of the Formative period. Races added to this list during the Classic period include Alazan, Perla, Pardo, Mochero, Chaparreño, Pagaladroga, Huayleño, Huancavelicano, Ancashino, and Piricinco. Only three more races appeared in the Postclassic period. These were Rienda, Chancayano, and Cuzco.

Five of the nineteen races that were postulated to have appeared by the Classic period have been identified on the ceramic maize specimens in this corpus. They include Confite Puneño, which has a Guatemalan counterpart Quicheño Rojo, Confite Morocho, Confite Puntiagudo, Mochero, and Chaparreño. Other races identified on the pottery that may be added to this list are Karapampa, Pollo, Pira, Güirua, Chococeño, Imbricado, Niñuelo, Oke, Choclero, and Chapalote. The Indians apparently had a wide variety of different kinds of maize during the Classic period.

Mexico

The plant remains from dry caves have provided evidence for a good evolutionary sequence of Mexican maize races. These include maize remains from Swallow Cave in the state of Chihuahua (Mangelsdorf and Lister 1956) and La Perra Cave in Tamaulipas dated ca. 2500 B.C. (Mangelsdorf, MacNeish, and Galinat 1956, 1967b). The remains from La Perra and Swallow Caves resembled tiny cobs of a very primitive pod-popcorn from Bat Cave, New Mexico (Mangelsdorf and Smith 1949; Mangelsdorf, Dick, and Cámara-Hernández 1967), but they were somewhat larger. The Swallow Cave maize from northwestern Mexico has been described as a prototype of the ancient race Chapalote, which is still grown in Mexico. The maize from La Perra Cave in northeastern Mexico was more closely related to the ancient race Nal Tel, which is still grown in Mexico and Guatemala. Nal Tel and Chapalote are closely related pod-popcorns, and the two races are often indistinguishable when one examines cob remains without kernels.

The most complete picture of the evolution of maize in Mexico has been provided by excavations in a series of five dry caves in the Valley of Tehuacán in southern Puebla and northern Oaxaca. Over 24,000 specimens of maize remains were obtained from these caves, and they provide a well-defined sequence showing how over a period of 6,500 years, maize evolved from a grass with tiny ears into a highly productive plant food. Maize is present at Tehuacán ca. 5000 B.C., according to Mangelsdorf et al. (1967a). Subsequent accelerator mass spectrometry (AMS) revised the dates to 3600 B.C. (Long et al. 1989). During the Abejas phase (originally dated ca. 3400–2300 B.C.), two new types of maize appeared. These include an early tripsacoid type and another type described as the race Nal Tel. These types are early agricultural maize. The ears are larger, and they exhibit more variability than the maize in earlier levels.

During the Palo Blanco phase at Teotihuacán, 200 B.C. to A.D. 800, a type of maize described as belonging to the Nal Tel–Chapalote complex predominated. During this period, a new type of slender popcorn resembling the Mexican race Arrocillo Amarillo and the Colombian race Pira Naranja was introduced. This is the antecedent maize of Pira/Nal Tel, the predominant type identified on Mexican and Peruvian vessels. Following the appearance of the new slender popcorn, five new races developed. These have been identified from the upper levels of Coxcatlán and El Riego Caves. They include the modern races Conico, Zapalote Chico, Tepecintle, Chalqueño, and an unidentified dent corn (Mangelsdorf, MacNeish, and Galinat 1967a, 178–200).

It is interesting that 63 percent of the depictions of maize on ceramic specimens from Oaxaca are of the races Pira/Nal Tel and Chapalote. The earliest, most primitive cultivated maize types in Mexico were still being represented on the pottery of Oaxaca roughly three thousand years after their first appearance and at a time when there is evidence that the Indians had access to more productive maize races. This raises the question of whether these were primitive, less productive types of maize being grown in Oaxaca during the Classic period or were other, more productive races grown for consumption and these primitive races maintained for ritual use. The answer may be provided in part by present-day peasants in the Valley of Oaxaca who continue to grow low-yielding, early maturing

Indian maize rather than opt for much higher yielding hybrid varieties available to them. Oaxacan peasants tend to plant just enough maize for their households, even when more land for surplus crops is available for planting and even in years when moisture conditions are favorable for increased production. The Indian peasant in Oaxaca chooses security over increased productivity. This same idea seems to influence the Indians of the Andean highlands, who still grow the primitive popcorn Confite Morocho although other, much higher yielding maize races could be grown there.

Races identified on Mexican specimens in this corpus that have also been identified in archaeological contexts include Nal Tel, Chapalote, Confite Puntiagudo/Canguil, and Pollo. Puya, Karapampa, and Oke have not been identified from any archaeological plant remains. Their presence on the pottery presumably represents the northward diffusion of these racial types from South America into Mesoamerica by A.D. 900.

Prehistoric Races of Maize and Cultural Historical Relationships

Maize has been the mainstay of subsistence for the Indians of Latin America for almost 4,000 years. From the time that agriculture began in the areas where New World civilizations developed, the most important crop was maize. This plant food provided the subsistence base for expanding populations: population growth, increased productivity of maize, and the development of civilization are intricately interwoven. Therefore, an understanding of the evolution of civilization in the New World is intimately tied to understanding the evolution of maize.

The findings from the ceramic maize replicas complement present knowledge from the archaeological record regarding the evolution of maize. By the Classic period, there was a wide variety of maize races in Peru, including some of the most productive types presently grown on the North Coast. Oaxaca apparently had fewer different types of maize. The main race seems to have been the primitive pod-popcorn Pira/Nal Tel, and the limited archaeological plant remains from Oaxaca appear to be Nal Tel (Ford 1976). There is also an indication that in Oaxaca during the Classic period, new races of maize were being introduced from South America.

If, indeed, new races were being introduced from South America, they were being transported into Mesoamerica as a result of cultural diffusion. This conclusion is inevitable because maize does not have a natural mechanism for seed dispersal and is therefore entirely dependent on humans for its survival. Maize races outside geographical areas of adaptation indicate cultural movements.

An understanding of cultural diffusion is vital to any study of sociopolitical systems. The questions of origins and histories of cultural groups, technologies, artistic styles, ideologies, trade routes, and patterns of exchange depend on whether there were cultural contacts linking geographical areas or whether cultures were evolving in isolation. Maize is a cultigen from which insights into cultural diffusion may be gained. There are cultural as well as botanical reasons why this is so.

In addition to its importance as a plant food, maize had symbolic and religious connotations that may be the reason for its frequent depiction. For example, maize is realistically depicted on numerous effigy vessels from Peru and Oaxaca in association with figures representing important deities or rulers. Votive offerings of maize were placed in shrines in ceremonial centers. Maize was a principal food placed in graves as offerings for the dead. In Peru, there is evidence that maize kernels were sometimes placed in the mouth and nostrils of the dead (Towle 1952a, 227). Maize served in gourd bowls and on plates, and occasionally in ceramic vessels, was buried with the dead as a food offering (Gumerman 1994). It is the most frequent food offering in burials on the North Coast of Peru and is found among people of different socioeconomic levels. Maize as a burial food offering was not reserved for rulers and high-status individuals. Chicha, a fermented drink made from sprouted maize kernels, was an important ritual beverage in Mexico and Peru (Oviedo y Valdés 1851–55).

In addition to its important uses as food and as an object of worship and veneration, certain varieties of maize had special medicinal uses (Cobo 1890–95). Ethnological evidence suggests that the use of maize for treating certain illnesses originated in pre-Columbian times. Among the Callahuayo Indians of the Bolivian highlands today, there is a group of traveling medicine men. Included among their various medicines is a type of podcorn. It is a primitive type of maize in which the individual kernels are completely enclosed by separate glumes or husks. This variety is almost extinct today and may have survived because of Indian beliefs in its curative powers. Podcorn, ground into a fine powder, is administered by the medicine men in the treatment of "el aire," a term that refers to conditions causing difficulty in breathing, including tuberculosis, asthma, and bronchitis (Cárdenas 1943). These Indian medicine men from the Andean highlands travel all over South America and into Central America to treat patients. They are thought to be practitioners of ancient medicine because along with their packs of remedies, they characteristically carry a *quena,* a reed flute of pre-Columbian origin (Cutler 1944, 293). These traveling medicine men are vectors of cultural diffusion.

I suggest that the various uses of different types of maize that are adapted to specific geographical areas and environmental conditions may have been an important factor in making maize a valuable exchange commodity, thereby linking distant cultures through trade. The sixteenth-century chronicler Bernal Díaz del Castillo reported that Indians traveled by canoe along the northern Pacific coast of South America and engaged in port-to-port trade in maize and salt (1934, 178). The Chibcha Indians of the Colombian highlands were engaged in extensive trade of salt and emeralds in exchange for gold (Bollaert 1860; Von Hagen 1952), and there was a widespread aboriginal salt trade in Mexico (Mendizabal 1928). Evidence presented here suggests that during the Classic period there was long-distance trade in maize and other perishable commodities that may have left few traces in the archaeological record.

This study represents an interdisciplinary approach combining botany and archaeology to bring new evidence to bear on anthropological problems. It demonstrates the usefulness of approaches based on interdisciplinary methodologies for testing hypotheses of cultural diffusion. With increasing sophistication and use of scientific techniques for improved recovery and more detailed analyses of archaeological remains, combined with more extensive excavations in intermediate areas, I propose that new evidence to show cultural contacts between Mesoamerica and the Andes will be brought to light and contribute to a greater understanding of the historical relationships between New World aboriginal cultures, thereby aiding understanding of cultural process.

Conclusions

Nineteen races of maize were identified on 129 vessels; sixteen on 72 Peruvian jars and ten on 56 Mexican specimens. This evidence indicates that during the Classic period, approximately A.D. 300 to A.D. 900, diverse maize types existed in Mesoamerica and northern South America.

Five of the sixteen races on Moche jars, including Confite Puneño, Confite Morocho, Confite Puntiagudo, Mochero, and Chaparreño, corroborate the hypothesis of Grobman et al. (1961) that these races had evolved by the end of the Classic period.

Six of the sixteen races identified on Peruvian jars are not found in Peru today. They include races from Chile, Bolivia, Ecuador, Colombia, Venezuela, and Mexico. This may indicate that these races, found today only outside Peru, were much more widely distributed in prehistory.

The identification of the race Choclero indicates that there may have been contact between the North Coast of Peru and the coast of Chile during the Classic period.

The identification of the Colombian races Pira, Pollo, Imbricado, and Güirua establishes the existence of these races by A.D. 900 and points to cultural relationships between highland Colombia and the North Coast of Peru during the Classic period.

Of the Colombian races identified, Güirua, Imbricado, and Pollo have counterparts in Central America. Güirua, presently found in Santa Marta, the area of development of the Tairona culture, has a counterpart in highland Guatemala. Its counterpart is the race Negro de Chimaltenango. The presence of the race Güirua in Peru by A.D. 900 suggests that there was contact between the two areas.

The identification of the Colombian race Pollo on Moche jars and Oaxacan urns shows that Pollo, which is believed to be closely related to the Mexican races Nal Tel and Chapalote, was present during the Classic period. The measurements on many of the vessels that have representations of Pollo are closer to the dimensions for Nal Tel and Chapalote. This seems to support the close relationship of these three primitive races as proposed by Mangelsdorf (1974). Comparative studies of archaeological remains identified as Pollo, Nal Tel, and Chapalote may demonstrate that these three races are derived from the same lineage and that Pollo is a manifestation of the southward diffusion of Nal Tel from central Mexico.

The identification of the highland races Karapampa from Bolivia and Oke from northwest Argentina on Peruvian and Mexican pottery demonstrates that these races were more widespread during the Classic period than they are today. Since they are not lowland races, their presence by A.D. 900 suggests that they were part of ancient exchange commodities being transported from the Andes Mountains down to the coast and northward into Mesoamerica.

The identification of Puya, a Venezuelan race found today in areas along the established Arawak trade routes that converged at the island of Trinidad, on a Oaxacan jar suggests that it was transported along ancient trade routes proposed by Chard (1950). Puya has a counterpart race in Colombia, but antecedent races of yellow dent corn are found in Mesoamerica, and it is believed that this is an example of diffusion southward from Mesoamerica into northern South America.

The races Nal Tel, Chapalote, and Pollo have been identified on Oaxacan urns and Peruvian jars. These races have been previously identified from archaeological maize remains for which the identifications here provide corroboration. Pira/Nal Tel is the race most frequently depicted on Zapotec urns and Moche jars in association with divine figures. This primitive popcorn may have had special symbolic significance for the Zapotec and Moche, and it was probably the most widely grown race during the Classic period. Although it is an ancient race and not as productive as other races that may have been available at that time, it has the advantage of being an early maturing variety that does not require a lot of moisture. This trait would have been particularly advantageous in Oaxaca, a relatively dry area, where rainfall is an important factor determining success of crop production, and on the dry desert coast of Peru. The maintenance of this primitive race may indicate that the people were more concerned with being assured of the success of their crop than with trying to grow more productive varieties of maize to provide a surplus.

Seed selection strategies among modern peoples may be a subject that merits further study before the introduction of modern maize hybrids leads to the extinction of indigenous Indian varieties and nullifies decision patterns based on them.

In addition to the identification of the Colombian race Pollo on a Oaxacan specimen, discussed above, Confite Puntiagudo/Canguil from Peru and Ecuador, respectively, has been identified on some mazorca sherds from Morelos which are dated ca. A.D. 900. These sherds indicate that connections between Mexico and South America were established by this time.

It would appear that all of the maize races identified on the pottery in this corpus had a wider geographical distribution during the Classic period than they do today. The North Coast of Peru may have been a major center of cultural exchange, connecting Mesoamerica and South America. The Classic period was characterized by tremendous activity and cultural development. This may have been attributable in part to active exchange between widespread cultural traditions.

The diffusion of different races of maize apparently gave rise to new races through hybridization and acclimatization to new environments. It is suggested that the expanded gene pool for maize races during the Classic period gave rise to more productive, more adaptive hybrids that contributed to the population increase during the Classic period.

There are some limitations regarding the identifications of maize races.

The measurements from the ceramic maize replicas are necessarily compared to data for living races. Evolution is a continuous process, and the maize races from a thousand years ago cannot be expected to match modern strains perfectly. In addition, modern distributions of maize races do not necessarily reflect prehistoric distributions. Furthermore, although the morphometric characters on which the identifications are based are those that are least affected by environmental parameters, there is always some question of variability within races and variability resulting from environmental factors. The number of specimens in this corpus bolsters the validity of the identifications. I must emphasize that though the racial identifications are based on observable, measurable characters, qualitative factors are also involved in the identifications, particularly when there are very similar alternative choices for identification. In cases when these similarities are repeated consistently with high frequency, both choices are considered to be equivalent types that have different names in different regions where they were collected and described. I further recognize that identifications based on morphology may not adequately reflect phylogenetic relationships as pointed out by John Doebley (1994). The purpose of this study was to make a thorough record of the botanical taxonomic information preserved in these unique terra-cotta casts and relate it wherever possible to the archaeological record and extant maize races. This body of data combined with molecular, phenetic, and paleoethnobotanical evidence will augment our understanding of the evolutionary history of maize.

This corpus includes a large number of Zapotec and Moche vessels depicting mold-made maize in museum collections in the United States, Mexico, Peru, Canada, Germany, Belgium, and France. This virtually exhaustive comparative study of these vessels adds weight by quantification to the validity of the identifications of the maize races, contributes to knowledge about the prehistoric distribution and evolution of these races, corroborates identifications of archaeological maize remains, and provides further evidence for cultural contacts between Andean and Mesoamerican cultures during the Classic period.

It has been said, "The more fully we understand the relationships of the various races and sub-races of maize in space and time, the more we will be able to infer about the civilizations and peoples who used and distributed this maize" (Anderson 1943, 63). This interdisciplinary study of the ceramic evidence for the spread of prehistoric races of maize contributes to achieving this goal.

APPENDIX A: FAKE ZAPOTEC URNS

As early as the 1860s fake urns were manufactured in Oaxaca and sold as authentic Zapotec urns (Boos 1966b, 15). Many of these urns were made using actual prehistoric molds and were so well executed that it was virtually impossible to determine whether they were bona fide antiquities or of recent manufacture. Until the development of thermoluminescence dating of Oaxaca pottery was advanced by David Zimmerman and applied in conjunction with an art historical stylistic study by Shaplin (1978), the stylistic diversity in Zapotec art was underestimated and earlier estimations of whether Zapotec urns were authentic or fake, either through stylistic analysis (Shaplin 1975) or identification of maize races on Zapotec urns (Eubanks 1973), were sometimes misguided. Two urns that Eubanks (1973) originally thought were suspect (figs. 98 and 123) because scientists believed the derivation of the types of maize represented was post-Columbian were subsequently determined by thermoluminescence dating to be prehistoric (Zimmerman personal communication February 25, August 4, 1975).

It was a significant contribution to determine that dent corn, the most productive corn grown in industrialized countries today, had in fact evolved at least a thousand years earlier than was thought and was present in Mexico by the pre-Columbian Classic period (Dunn 1977).

There were some pieces at the Museo Nacional de Antropología, the Museum für Völkerkunde in Berlin, and the Musées Royaux d'Art et d'Histoire in Brussels that were determined to be forgeries. The pieces at the latter two European museums were dated by thermoluminescence. These urns are presented here in catalog format with the letter F prefix to distinguish them as fakes.

F-1. Fig. 127. Urn depicting the God Cocijo.
Museum für Völkerkunde Berlin 31104.
Provenience: Santa Maria Sola, Oaxaca.
Fabric: dark gray with traces of red.
Dimensions: ht. 38.5 cm, width 13.3 cm.
Race of maize: Nal Tel/Pira.

The figure is seated cross-legged with palms on knees. The posture is typical for Zapotec urns, but the depiction of five fingers and toes is unusual. The figure wears a Cocijo mask with nose knot, bifurcated tongue, and eye and cheek elements. It has a feather cape with pectoral glyph, elaborate earspools, plain bracelets on both wrists, and a loincloth with three tassels and an undulating motif. The headdress has five standing ears of maize backed by a *penache*. This urn is illustrated by Schuler-Schömig (1970, number 43) and was recognized by Shaplin as early as 1978 to be a fake (Philippa D. Shaplin personal communication, February 10, 1978).

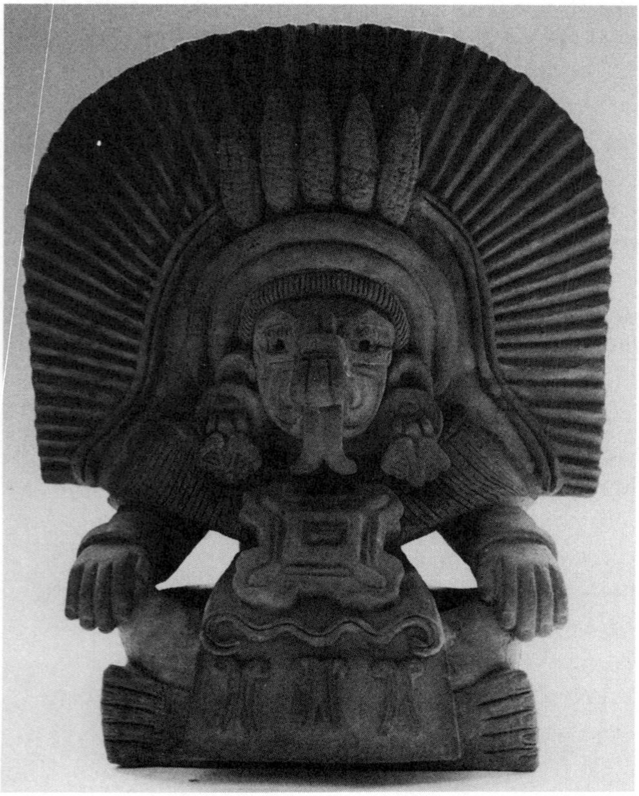

Fig. 127. F-1. Urn depicting the God Cocijo.

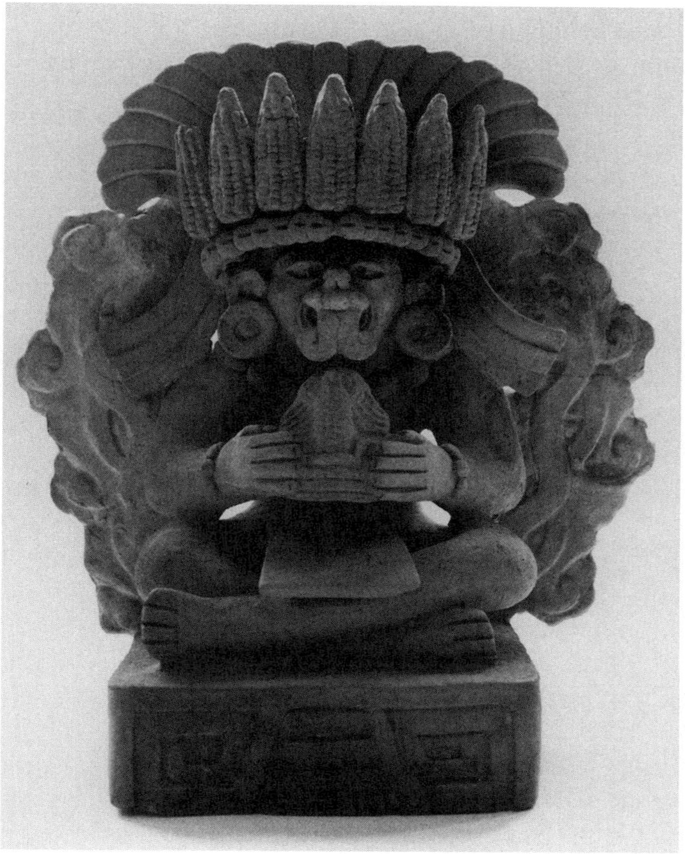

Fig. 128. F-2. Urn depicting the God Cocijo.

F-2. Fig. 128. Urn depicting the God Cocijo.
Museum für Völkerkunde Berlin 35275.
Provenience: Talpan.
Fabric: gray.
Dimensions: ht. 37.5 cm, width 36.3 cm.
Race of maize: Nal Tel/Pira.

The figure is seated cross-legged and holds an ear of maize with staminate tip and stylized silks. The posture is typical for Zapotec urns, but the depicting of five fingers is unusual. The figure wears a Cocijo mask with nose knot, bifurcated tongue, and eye and cheek elements. It has large, round earspools, bracelets on both wrists, a bead necklace, and an unadorned loincloth. It sits on a pedestal whose base is incised with the hill symbol and is backed by an elaborate dossel. The headdress has seven standing ears of maize surmounted by a *penache*.

This urn is illustrated by Schuler-Schömig (1970, number 115) and was recognized by Shaplin as early as 1978 to be a fake (Philippa D. Shaplin personal communication, February 10, 1978).

F-3. Fig. 129. Urn depicting the God with maize in his headdress.
Museum für Völkerkunde Berlin 39742.
Provenience: unknown.
Fabric: gray.
Dimensions: ht. 33.0 cm, width 12.1 cm.
Race of maize: Nal Tel/Pira.

The figure is seated cross-legged with palms on knees. The posture is typical for Zapotec urns, but the depiction of five fingers is unusual. The figure does not wear a mask but has a feather cape with pectoral glyph, plain earspools, bracelets on both wrists, and an undecorated loincloth. The

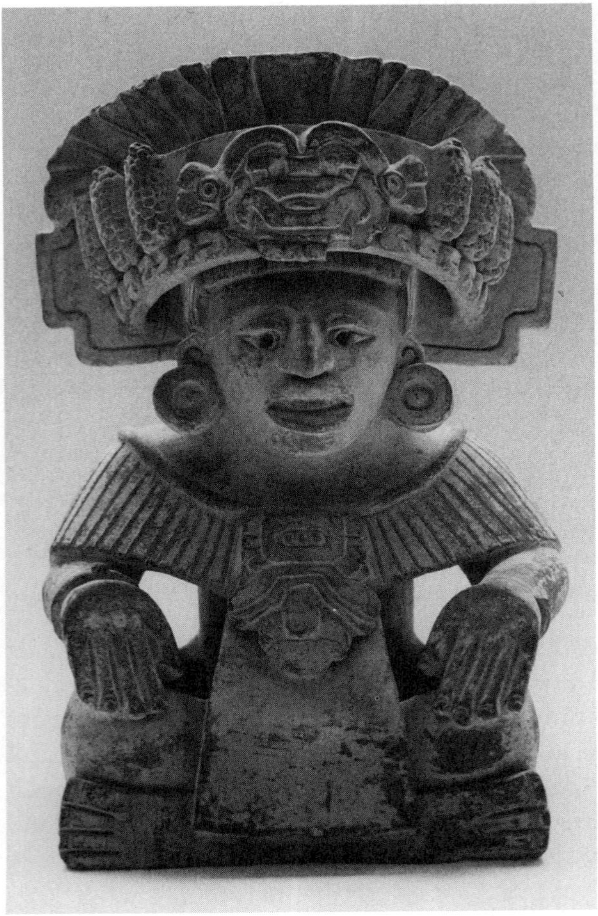

Fig. 129. F-3. Urn depicting the God with maize in his headdress.

headdress has a central "C" glyph flanked on either side by three standing ears of maize surmounted by a *penache*. This urn is illustrated by Schuler-Schömig (1970, number 159).

F-4. Fig. 130. Urn depicting the Bat God.
Museum für Völkerkunde Berlin 35283.
Provenience: Talpan.
Fabric: gray.
Dimensions: ht. 45.0 cm, width 27.5 cm.
Race of maize: Nal Tel/Pira.

This figure of the Bat God is standing upright on clawed feet; its clawed arms are held up at either side of its body. It has large bat-like ears, fangs, and eye and cheek elements. It wears a plain loincloth, a cord necklace with pectoral ornament and bells, and round earspools from which mold-made maize ears are suspended. The back of the figure is adorned with a

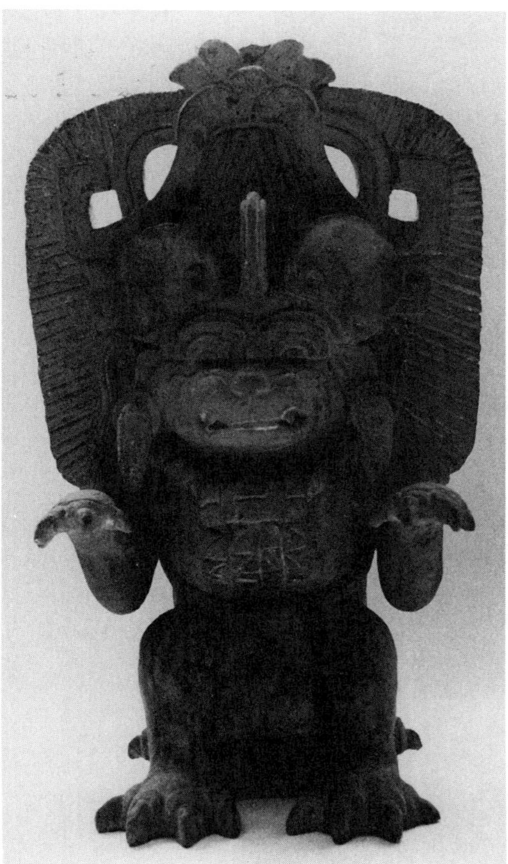

Fig. 130. F-4. Urn depicting the Bat God.

large *penache* that extends upward from its shoulders to above its head. This urn is illustrated by Schuler-Schömig (1970, number 70).

F-5. Fig. 131. Urn depicting the God with maize in his headdress.
Musées Royaux d'Art et d'Histoire 158878B.
Provenience: unknown.
Fabric: gray.
Race of maize: unidentified.

The figure is kneeling and holding an ear of maize. The posture is atypical, and the depiction of five fingers is unusual. The figure does not wear a mask but has three beads across the bridge of its nose. It has plain round earspools, bracelets on both wrists, and a loincloth or skirt with an incised border design of alternating vertical and horizontal parallel lines. The headdress has a central "C" glyph surmounted by two large plumes that terminate in snake heads and a *penache*.

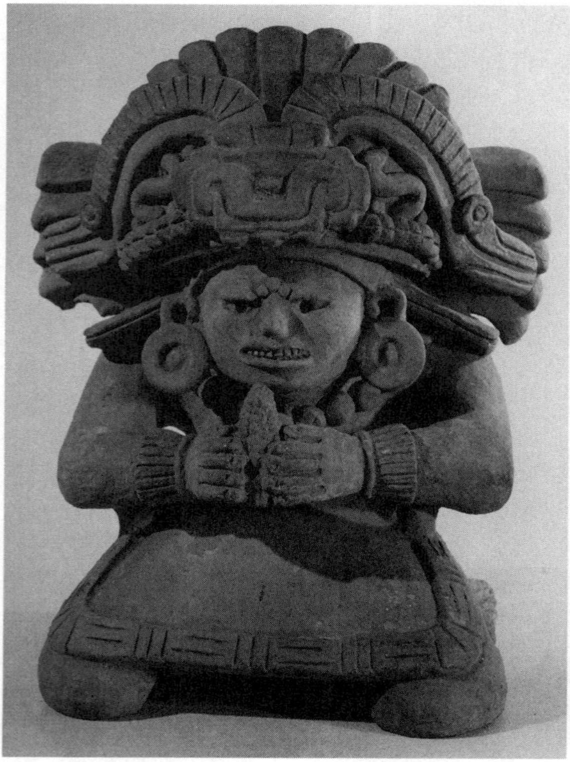

Fig. 131. F-5. Urn depicting the God with maize in his headdress.

F-6. Fig. 132. Urn encircled with ears of maize.
Royal Ontario Museum HM-1947.
Provenience: unknown.
Dimensions: ht. 15.5 cm, width 24.5 cm.
Period: Monte Albán V.
Race of maize: unidentified.

This is a tripod vessel encircled by twenty-three ears of maize. There are dog-like heads on the tripod legs. The race is unidentified because measurements were unavailable.

This urn was dated by David W. Zimmerman using thermoluminescence and found to be a modern forgery. "The specimen (sample no. TOAX 25) was 14 rads 'equivalent dose,' i.e., 45–50 years old" (Philippa D. Shaplin personal communication, January 12, 1978).

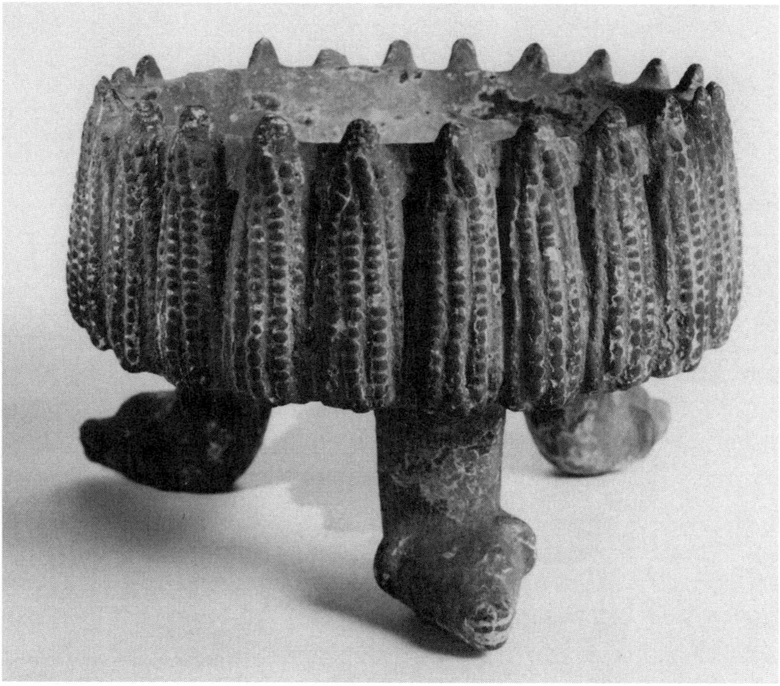

Fig. 132. F-6. Urn encircled with ears of maize.

F-7. Fig. 133. Fragment of an urn depicting the God of Glyph "L."
Museo Nacional de Antropología.
Provenience: Monte Albán.
Fabric: gray.
Dimensions: ht. 16.9 cm, width 11.3 cm.
Race of maize: unidentified.

This figure is not altogether unlike some of those that fit the category the God of Glyph "L." This piece has been badly damaged, an obvious ruse of a forger. The treatment of the mouth gives this piece away as a fake. The face is not characteristic of any Zapotec ceramics. The headdress contains two nicely molded sections of maize ears.

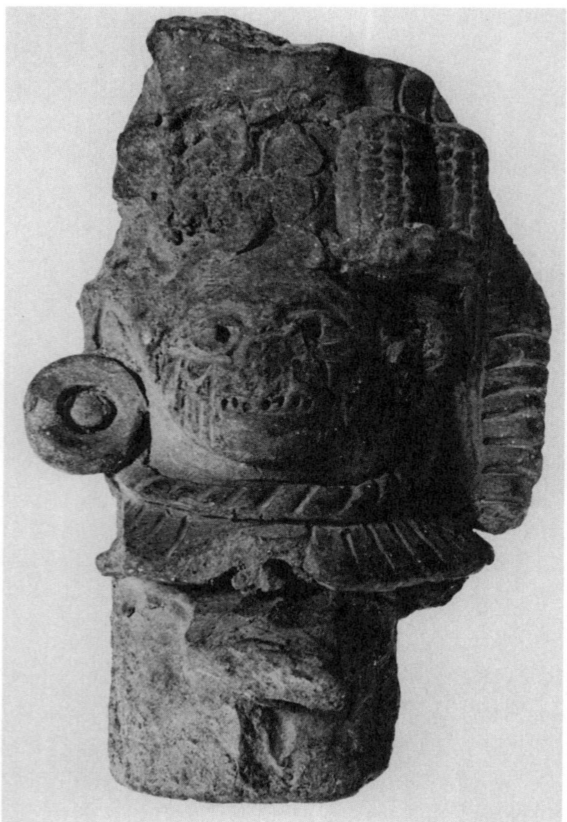

Fig. 133. F-7. Fragment of an urn depicting the God of Glyph "L."

F-8. Fig. 134. Fragment of an urn depicting the God of Glyph "L."
Museo Nacional de Antropología B128.
Provenience: unknown.

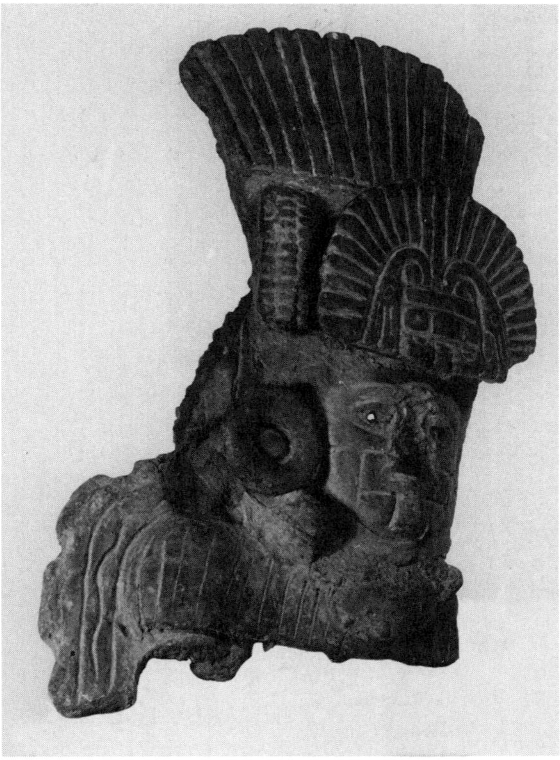

Fig. 134. F-8. Fragment of an urn depicting the God of Glyph "L."

Fabric: brown.
Dimensions: ht. 25.0 cm, width 15.6 cm.
Race of maize: unidentified.

Figure 134 illustrates another forgery of the God of Glyph "L" that has been purposely damaged to enhance the deception that it is a prehistoric artifact. The figure has a glyph that is a poor attempt to imitate a Zapotec "C" glyph as the central medallion of the headdress. The ear of maize on the side of the glyph is mold-made but is inverted. This is a positively identified example of a fake with an upside-down maize ear. It seems illogical that people who worshiped maize would purposely depict it this way. The figure wears a feather cape and large round earspools. A decorative panel with an incised wavy line extends from its shoulders.

F-9. Fig. 135. Fragment of an urn depicting the God Cocijo.
Museo Nacional de Antropología 6–366.

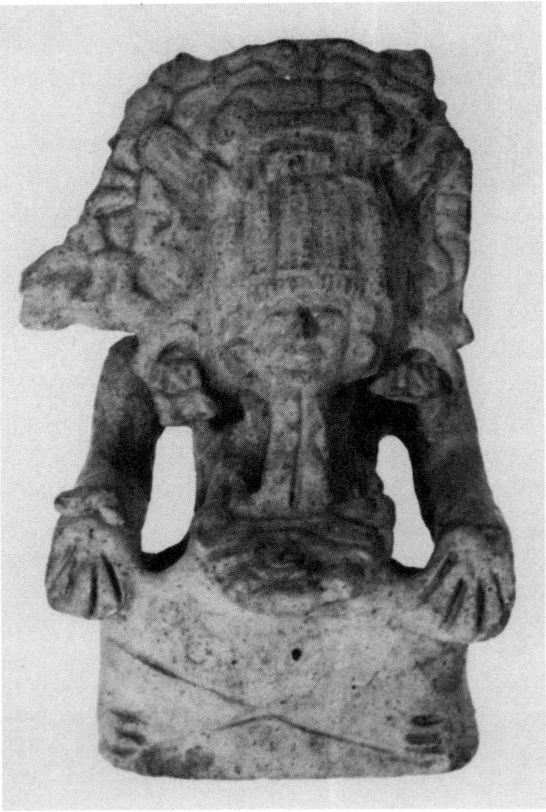

Fig. 135. F-9. Fragment of an urn depicting the God Cocijo.

Provenience: unknown.
Fabric: yellow.
Dimensions: ht. 24.4 cm, width 16.3 cm.
Race of maize: unidentified.

This piece is poorly executed, and the details are not clear. In all, there are seven ears of maize in the headdress, but it is impossible to tell for certain if they are from a mold. Certain numbers have significance in Mesoamerican numerology, but seven does not (Donald L. Brockington personal communication). The figure has an exaggeratedly long bifurcated tongue hanging from its chin but no other Cocijo mask elements. In addition, the clay out of which this urn is fashioned is yellow, and I have not seen any other urn constructed out of clay like this.

APPENDIX B: RACES OF MAIZE FROM LATIN AMERICA

Race	Ear length, cm	Ear diameter, cm	Kernel width, mm	Kernel thickness, mm	Row number
PALOMERO TOLUQUENO	10.2	3.4	4.7	2.8	23
ARROCILLO	9.8	2.7	5.5	2.5	15.4
CHAPALOTE	11	2.9	6.7	4.1	12.3
NAL TEL MEXICO	7.9	2.7	6.7	3.9	11.4
CACAHUACINTLE	14.5	4.7	9.8	5.2	15.2
HARINOSO DE OCHO	19.1	3.8	12	4.4	8
ELOTES	17.1	4	11	4.5	9.9
OLOTON MEXICO	18.3	4.3	9.7	6	11.7
MAIZ DULCE	13.7	4.5	8.5	4	14.5
CONICO MEXICO	13.6	4.3	6.6	3.6	16
REVENTADOR	16.5	3.2	7.4	3.6	11.9
TABLONCILLO	16.4	4.1	12	4.3	9.1
TAB PERLA	17	3.7	11	4.2	8.3
TEHUA MEXICO	19.1	5.7	9.1	3.9	17
TEPECINTLE	10.4	4.9	9.1	3.7	11.8
COMITECO	28.7	5.2	9.5	4.5	13.5
JALA MEXICO	30.5	5.9	11	4.6	14.7
ZAPALOTE CHICO	9.9	4.2	9.8	3.6	10.7
ZAPALOTE GRANDE	14.8	4.9	9.3	3.8	15.7
PEPITILLA	12.3	5.3	7.9	3.5	15.5
OLOTILLO	19.8	3.8	11	3.9	9.4

(continued)

Race	Ear length, cm	Ear diameter, cm	Kernel width, mm	Kernel thickness, mm	Row number
TUXPENO	19.7	4.4	9.3	3.7	12.6
VANDENO	17.2	5.1	9.1	3.6	13.2
CHALQUENO	16	4.9	7.2	3.9	16.6
CELAYA MEXICO	17	4.5	9.1	3.9	12.4
CONICO NORTENO	13.1	4.6	7.3	3.5	16
BOLITA MEXICO	11.6	4.2	10	4.1	10.2
WHITE POP	11.4	3.7	7.5	3.5	13.7
YELLOW POP	11.1	3	6.1	3.7	14.1
WHITE DENT	11	3.7	9.3	4.5	10.2
CANILLA CUBA	20.6	3.8	7.8	4.2	13.4
TUSON CUBA	18.1	5.2	9.3	3.8	15.4
CRIOLLO CUBA	19	4.9	9.1	4.1	14.9
ARGENTINA CUBA	17.2	4.1	8.7	4	13
POLLO COLOMBIA	8.8	2.9	7.9	5	10
PIRA COLOMBIA	11.2	2.6	6.5	3.6	11
PIRA NARANJA	15.7	2.8	5.5	3.8	14.7
CLAVO COLOMBIA	20	3.2	8.3	5.2	11.1
GUIRUA COLOMBIA	16.2	4.3	8	4.9	12.2
CARIACO COLOMBIA	11.9	4.8	8.8	4.7	15.8
ANDAQUI COLOMBIA	15	3.6	9.6	4.6	10.3
IMBRICADO	11	3.9	7.2	5.3	14.3
SABANERO COLOMBIA	14	4.4	11	6.1	11.7
CABUYA COLOMBIA	16	3.7	11	6.4	9.3
MONTANA COLOMBIA	21.8	4.8	11	6.2	13.2
CAPIO COLOMBIA	19.8	5.1	11	7	13

AMAGACENO	18.4	4.2	9.6	5.3	12.6
COMUN COLOMBIA	18.2	4.8	11	4.9	12.7
YUCATAN COLOMBIA	17.5	4.6	11	4.2	10.8
CACAO COLOMBIA	15	4.6	10	5.3	12.6
COSTENO COLOMBIA	15.5	4.7	9.9	4.4	13.1
NEGRITO COLOMBIA	13.1	4.3	8.4	4.3	14.4
PUYA COLOMBIA	17.2	3.8	9	4.2	11.4
PUYA GRANDE COLOMBIA	18.7	3.9	9.5	4.4	11.4
CHOCOCENO COLOMBIA	11	3.7	6.5	4.1	17.2
AVATI PICHINGA	15.2	2.7	5.4	3.6	13.8
MOROTI MITA	15	3.1	8.5	4.7	10.2
MOROTI	17.8	4	8.7	4.8	12.7
MOROTI TI	19.2	4.1	8.2	5.1	19.2
DJAKAIRA	18.7	3.9	8.5	4.8	14
MOROTI GUAPI	13	4.6	8.7	4.7	14.8
CAINGANG	19.6	4.2	9.8	4.6	12.3
LENHA BRAZIL	18.5	5.9	8	4	22.5
CAPIA AMARILLO	11.9	4.6	9.1	4.3	12.7
CAPIA BLANCO	12.3	4.9	9	4.3	14.1
CAPIA A. DE OCHO	12.8	3.8	9.5	4	9.7
CHULPI ARGENTINA	10.4	4.5	7.4	3.9	15.5
OKE ARGENTINA	8.2	3.4	7.9	4.3	10.7
ACHILLI ARGENTINA	15.5	4.5	10	4.1	10
MARRON ARGENTINA	13.2	4.5	8.4	4.2	12
MOROCHO ARGENTINA	13.4	4.5	8.2	4.2	14.5
ALTIPLANO ARGENTINA	6.1	3.6	8	4.9	12.2
BOLA BLANCO	8.1	3.5	6.3	4.9	15.2

(continued)

Race	Ear length, cm	Ear diameter, cm	Kernel width, mm	Kernel thickness, mm	Row number
CULLI ARGENTINA	11.9	4.1	9.1	4.4	10.5
AMARILLO DE OCHO	13.8	3.7	9.7	4	8
PISINCHO ARGENTINA	11.1	3.7	5.4	3.7	18
CRISTAL PARAGUAY	17.4	4.1	8.8	4.6	13.2
CRISTAL SANTA CRUZ	18.4	4.8	8.8	4.8	15.6
CRISTAL REYES	20.8	3.8	8	5.4	14.8
CANARIO DE OCHO	17.2	3.3	11	4.3	8
CATETO ORANGE	18	3.9	9	4.2	11.8
NORTHERN CATETO	19.4	5.4	9.1	4.1	15.8
INTERLOCKED SOFT	25.6	3.2	8.6	5.1	12.1
CONFITE PUNENO	5.8	3.4	6.7	5.3	14
ALITIPLANO BOLIVIA	8.6	4.2	8	5.1	12.2
PATILLO BOLIVIA	7.5	3.7	7.2	4.8	14
KCELLO BOLIVIA	11.9	3.7	9.5	4.1	10
KULLI BOLIVIA	9.2	4.4	8.4	5	12.3
HUILCAPARU	15.7	4.8	9.9	4.4	12.2
CHAKE-SARA	12.5	3.7	8.5	4.1	11.4
AYSUMA	10.8	3.6	8.7	4.4	9.5
PATILLO GRANDE	14.2	4.3	9.1	4.2	12.6
CHECCHI BOLIVIA	10.6	4.3	9	4.2	13.5
CUZCO-HUILICAPARU	12.2	4.8	11	4.5	9.2
PARU BOLIVIA	9.8	5.7	7.6	5	18.1
CHUSPILLU	11.7	5.1	6.8	3.8	19.7
CUZCO BOLIVIANA	14.4	5.3	13	5.1	8.7
PISANKALLA	12.1	3.3	5.4	3.7	16.5

UCHUQUILLA BOLIVIA	15	3.6	11	3.9	8.6
KARAPAMPA	12.2	3	7.9	3.3	8.8
ARGENTINO BOLIVIA	17.9	4.3	8.9	4.2	10.4
NINUELO BOLIVIA	12.1	3.2	9.5	4.3	8.2
CAMBA BOLIVIA	18.1	4.1	8.8	5.1	14.5
MORADO BOLIVIA	16.6	3.5	7.7	4.8	13.5
PEROLA BOLIVIA	16	3.4	8	4.3	12.9
YUNQUENO BOLIVIA	18.4	4.3	9.1	4.5	13.4
POJOSO CHICO BOLIVIA	15	3.9	7	4.7	14.2
CHOLITO BOLIVIA	13.5	4.2	7.7	4.2	16.2
CUBANO DENTADO	15.6	4.5	9.3	4.2	13.3
CATETO BOLIVIA	17	3.5	8.1	4.2	12.9
PORORO BOLIVIA	12.3	2.7	5	3.3	15
COROICO BLANCO	20.3	3.9	7.7	4.9	16.4
COROICO AMARILLO	23.1	3.8	8.9	5.5	12.9
COROICO	26.4	3.7	9.7	5.5	11.2
ENANO BOLIVIA	6.9	2.7	4.8	3.3	16.4
MARCAME	10.9	5	8.7	5.4	15.4
POLULO CHILE	9.6	2.4	5.3	3.5	15.4
NEGRITO CHILENO	6.8	4	8.4	4.7	17
CHULPI CHILE	10.1	4.7	7.4	4.5	17.4
CAPIO CHICO CHILENO	8.6	4.5	7.3	4.5	17.2
CAPIO GRANDE CHILE	12.2	5.3	6.3	3.9	19.1
CAPIO NEGRO CHILE	12.7	3.4	8	4.4	18.5
CHUTUCUNO CHICO	8.3	3.2	4.8	3.4	19.6
CHUTUCUNO GRANDE	12.7	4	6.7	4	17.6
HARINOSO TARAPAQUENO	14.8	5.4	11	4.5	12.4

(*continued*)

Race	Ear length, cm	Ear diameter, cm	Kernel width, mm	Kernel thickness, mm	Row number
CHOCLERO CHILE	14.9	5.6	7.1	4.1	22.1
CAMELIA CHILE	19.1	4.5	8.3	4.3	14.3
CURAGUA	13.9	2.9	6.2	3.8	18.8
CURAGUA GRANDE	18	4	6.4	4.9	19
CRISTAL CHILENO	19.7	4.3	9.1	4.6	13.6
DENTADO COMERCIAL	20.3	5.1	10	4.8	14.4
ARAUCANO CHILE	10.4	2.8	7.8	5	9.7
CRISTAL NORTENO	23.1	3.5	11	4.4	8
DULCE EVERGREEN	20.5	4.9	8.5	3.9	16.4
DULCE GOLD BANTAM	18.2	4.3	9.3	4	13.4
CANGUIL ECUADOR	11.1	3.3	5.6	4.5	15
SUB-RAZA GRUESO	11.9	3.9	4.2	4.4	16.8
SABANERO ECUADOR	13.9	4.6	10	6	12.7
CUZCO ECUADOR	15.1	5	13	5.5	9.7
MISHCA ECUADOR	11.5	4.2	5.9	5.3	10.2
PATILLO ECUADOR	9.9	3.6	7.6	5.6	10
RACIMO DE UVA	12.2	4.5	9.4	5.7	12
KCELLO ECUADOR	12.6	3.8	10	5.5	9.2
CHILLO ECUADOR	15.9	5.9	13	6.3	12
CHULPI ECUADOR	12.6	4.7	7.9	5.1	16.3
MOROCHON ECUADOR	16.7	4.4	11	5.9	10.7
HUANDANGO	19.9	3.8	11	6	8.3
MONTANA ECUADOR	26.6	4.2	11	5.2	11.9
BLANCO HARINOSO	16.3	5.2	11	5.4	12.4
CONICO DENTADO	17.4	4.6	13	3.2	12.5
UCHIMA ECUADOR	16.7	4.5	9.8	4.9	12.8

CLAVITO ECUADOR	13.1	3.1	8.4	5	10.1
POJOSO CHICO ECUADOR	13.7	3.9	7.3	4.8	16.4
TUSILLA ECUADOR	16.3	3.3	7.7	4.5	12.4
GALLINA ECUADOR	17.5	4.6	8.8	4.5	15.4
CANDELA ECUADOR	15.6	4.2	8.1	4.2	14.3
CHOCOCENO ECUADOR	12.4	4.1	6.6	4.2	19.6
CHIRIMITO VENEZUELA	9.3	3	6.7	3.6	13.1
ARAGUITO VENEZUELA	9.2	3.1	9.7	3.7	13.2
POLLO VENEZUELA	11.2	3.7	9.9	5.2	9.8
CANILLA VENEZUELA	16.7	3.4	6.7	4	14.6
PIRA VENEZUELA	10.9	2.6	5.8	3	10.8
CARIACO VENEZUELA	13.3	4.3	7.6	3.9	15
GUARIBERO	11.7	3.6	7	3.8	15.5
SABANERO VENEZUELA	16	4.4	9.4	5.7	12.3
HUEVITO VENEZUELA	15.6	3.8	9.8	4.9	10.5
PUYA VENEZUELA	17.1	3.8	7.1	3.4	13.6
TUSON VENEZUELA	17	4.8	9.1	4.1	14.9
CUBA YELLOW VENEZUELA	16.7	4	8.6	4.1	13.1
CHANDELLE VENEZUELA	18.2	3.6	8.1	3.9	11.6
COSTENO VENEZUELA	16	4.5	9	4.1	13.7
PUYA GRANDE VENEZUELA	21.6	4.6	9	4.2	13.6
TUXPENO VENEZUELA	19.2	4.9	9.5	3.9	13.3
COMUN VENEZUELA	18.4	4.1	9.3	4.1	11.7
NEGRITO VENEZUELA	16.1	4.1	8.4	4.3	13.7
CACAO VENEZUELA	15	4.6	10	5.3	12.6
CONFITE MOROCHO	6.8	2.3	5.9	4.6	10.7
CON PUNTIAGADO	8.9	3.2	5.6	4.3	15.1

(continued)

Race	Ear length, cm	Ear diameter, cm	Kernel width, mm	Kernel thickness, mm	Row number
KCULLI PERU	9.2	4.7	10	6	12
CONFITE PUNENO	6.6	3.5	6.7	5	16
ENANO PERU	7.2	2.5	5.3	3.7	16
HUAYLENO	11.6	5	7.9	5.3	8.6
CHULLPI PERU	8.6	4.6	8	4.3	18
GRANADA PERU	8.6	5.2	9.7	6	12
PARO PERU	11.5	5.8	8.9	5.1	14.7
MOROCHO PERU	12.2	4.3	9.8	5.7	10.8
HUANCAVELICANO PER	11.5	4.5	12	6.2	8.8
MOCHERO PERU	7.6	4.1	7.6	4.5	14
PAGALADROGA PERU	14.3	4.2	7.8	4.1	15
CHAPARRENO	10.5	4.2	8.4	4.4	14
RABO DE ZORRO	18.5	4	7.9	5.7	10.4
PIRICINCO	24	3.7	7.8	5.1	10
ANCASHINO	13.3	5.4	8.4	5.3	13.8
SHAJATU	10.3	4.8	7.8	5	16.3
ALAZAN PERU	15.8	4.9	9.2	4.8	14.2
SABANERO PERU	13.7	4.6	12	5	10.1
UCHUQUILLA	16.7	4.5	11	4.4	9.5
CUZCO CRISTAL AMAR	13.2	4.9	13	5.2	8.4
CUZCO PERU	12.8	4.2	13	5.4	8.3
PISCCORUNTO	11.8	5.2	11	5.2	12
AREQUIPENO	13	5.2	12	4.5	10.7
HUACHANO	14.7	4.6	11	4.5	11
CHANCAYANO	14.8	5.3	11	4.8	11.7
SG HUANCAVELICANO	8.1	4.7	12	5.5	11.1

PERLA PERU	18.4	5	8.8	5	15.6
RIENDA PERU	16.9	3.5	7.2	4.6	14
MARANON PERU	16.7	5	8.5	5.6	15
CHIMLOS PERU	22.7	3.6	7.5	5.6	14
CUZCO GIGANTE PERU	14.9	5.6	16	6	8.3
PARDO PERU	19.7	4.2	12	4.5	7.8
ALEMAN PERU	13.1	4.5	8.6	5	13
CHUNCHO PERU	17.8	5.4	8.5	4.9	15.2
ARIZONA PERU	17.4	5.2	9.7	5.1	14.2
JORA PERU	17.4	5.1	9	4.5	14.5
CORUCA	14.1	6.1	12	5	10.6
MORO CAJABAMBINO	14.2	3.6	9.3	5.7	10.7
MORADO CANTENO	12.9	4.7	10	4.8	11
PERLILLA PERU	11.1	4.2	6.6	5	16.6
CUBAN FLINT	17.4	4.4	8.8	3.8	14.2
HAITIAN YELLLO	18	4.1	9.4	4.5	11.6
COAST TROP FLINT	21	4.6	8.7	4.4	14.7
MAIZ CHANDELLE	19.6	3.3	7	4	12.5
EARLY CARIBBEAN	18	4.1	9.4	4.4	11.6
ST CROIX	22.9	5.1	11	4.6	12.4
TUSON CARIBBEAN	21.6	5.2	10	4.2	13.4
NAL TEL AM BAJA	11.2	3.4	8.3	4.1	11.7
NAL TEL BLAN BAJA	13	3.4	8.4	3.9	12
NAL TEL AM ALTA	14	3.7	8	5.5	12
NAL TEL BLAN ALTA	13	3.4	7.8	4.7	12
NAL TEL OCHO	15	4.1	10	4.8	11
IMBRICADO CA	16.2	4.8	8.6	4.6	14.4
SERRANO	12.8	3.4	10	5.2	10.5

(continued)

Race	Ear length/cm	Ear diameter/cm	Kernel width/mm	Kernel thickness/mm	Row number
SAN MARCENO	16.5	4.1	11	5.7	9.9
QUICHENO ALL	13.8	3.6	8.7	6.8	12.5
QUICHENO EARLY	13.5	3.6	8.2	4.9	13
QUICHENO LATE	14	3.6	8.4	5.8	12.3
QUICHENO ROJO	13.1	3.4	7	4.8	13
QUICHENO GRUESO	12	7.1	8	6	14
ND CHIMALTENANGO	15.8	3.9	9.9	5.2	10.3
ND TIERRA FRIA	14.1	4.1	10	5.2	10.3
ND TIERRA CALIENTE	14.1	4.1	9.2	4.2	11.5
SALPOR CENTRAL AMERICA	19.7	5.7	12	5.9	13
SALPOR TARDIO	15.8	4.6	11	6.2	11
OLOTON CENTRAL AMERICA	21.1	4.5	9.7	5.5	12.6
COMITECO CENTRAL AMERICA	20.1	4.9	10	5.1	13.1
DZIT-BACAL	17.2	3.8	10	4.3	9.2
TEPECINTLE CENTRAL AMERICA	18.4	4.7	9.1	4.5	14
TUXPENO CENTRAL AMERICA	17.8	4.9	9.3	4.2	14

APPENDIX C: TOP FIVE RACIAL IDENTIFICATIONS FOR CERAMIC MAIZE MODELS

Specimen	Race 1	Race 2	Race 3	Race 4	Race 5
Z-1	Karapampa	Reventador	Oke	Nal Tel Amarillo	Nal Tel Blanco
Z-2	Confite Morocho	Pira Venezuela	Avati	Pira	Polulo
Z-3	Confite Morocho	Pira Venezuela	Avati	Pira	Polulo
Z-4	Confite Morocho	Avati	Pira Venezuela	Polulo	Pira
Z-5	Araucano	Pollo	Confite Morocho	Nal Tel	Pira
Z-6	Pira	Confite Morocho	Nal Tel	Pira Venezuela	Chapalote
Z-7	Pira	Nal Tel	Karapampa	Reventador	Chapalote
Z-8	Nal Tel	Pira	Confite Morocho	Avati	Chapalote
Z-10	Nal Tel	Pira	Araucano	Pollo	Chapalote
Z-11	Nal Tel	Pira Ven	Pollo	Chapalote	Moroti
Z-12	Araucano	Confite Morocho	Pollo	Nal Tel	Pira
Z-13	Confite Morocho	Pira	Pira Venezuela	Nal Tel	Avati
Z-14	Pira	Nal Tel	Karapampa	Chapalote	Pira Ven
Z-15	Confite Morocho	Pira Ven	Pira	Nal Tel	Avati
Z-16	Araucano	Pollo	Clavito	Chapalote	Nal Tel
Z-17	Confite Morocho	Pira	Pira Venezuela	Nal Tel	Avati
Z-18	Confite Morocho	Pira	Nal Tel	Pira Venezuela	Chapalote
Z-19	Confite Morocho	Pira	Nal Tel	Chapalote	Pira Venezuela
Z-20	Pira	Confite Morocho	Nal Tel	Pira Ven	Chapalote
Z-21	Pira	Pira Venezuela	Nal Tel	Confite Morocho	Chapalote
Z-22	Pira	Nal Tel	Pira Venezuela	Karapampa	Chapalote

(*continued*)

Specimen	Race 1	Race 2	Race 3	Race 4	Race 5
Z-23	Confite Morocho	Pira Venezuela	Pira	Nal Tel	Polulo
Z-24	Oke	Nal Tel Amarillo	Altiplano	White pop	Patillo
Z-25	Pira	Nal Tel	Araucano	Karapampa	Pollo
Z-26	Karapampa	Moroti	Ninuelo	Reventador	Nal Tel
Z-28	Pira Venezuela	Pira	Nal Tel	Polulo	Confite Morocho
Z-29	Pira Venezuela	Pira Venezuela	Pira	Nal Tel	Avati
Z-31	Pira	Confite Morocho	Nal Tel	Chapalote	Pira Venezuela
Z-32	Pira	Confite Morocho	Nal Tel	Pira Ven	Avati
Z-33	Pira	Nal Tel	Pira Venezuela	Chapalote	Avati
Z-34	Confite Morocho	Pira Ven	Pira	Nal Tel	Polulo
Z-35	Confite Morocho	Pira	Araucano	Nal Tel	Pollo
Z-36	Pisankalla	Confite Puntiagudo	Canguil	Pisincho	Chutucuno
Z-38	Confite Morocho	Nal Tel	Enano Peru	Pira	Polulo
Z-39	Nal Tel	Pira	Chapalote	Chirimito	Chandelle
Z-41	Confite Morocho	Oke	White dent	Ninuelo	Nal Tel Amarillo
Z-42	Confite Morocho	Pira Venezuela	Pira	Nal Tel	Avati
Z-43	Puya Ven	Pojoso Chico	Canilla	Chandelle	Guaribero
Z-44	Ninuela	Canario	Karapampa	Araucano	Moroti
Z-45	Chapalote	Pollo	Araucano	Nal Tel	Nal Tel Blanco
Z-46	Oke	Aysuma	Piricinco	Moroti	Nal Tel Amarillo
Z-48	Confite Morocho	Nal Tel	Pira	Chapalote	Araucano
Z-49	Confite Morocho	Pira Venezuela	Nal Tel	Pira	Polulo
Z-51	Pira	Nal Tel	Chapalote	Yellow pop	Chirimito
Z-52	Curagua	Polulo	Capio Negro	Pira Naranja	Chapalote
Z-53	Pira	Chapalote	Chirimito	Nal Tel	Yellow pop
Z-54	Confite Morocho	Pira Venezuela	Pira	Nal Tel	Polulo
Z-56	Chapalote	Chandelle	Tusilla	Canilla	Reventador
Z-57	Canguil	Confite Puneño	Bola Blanco	Confite Puntiagudo	Confite Puneño

P-1	Confite Morocho	Pira	Nal Tel	Pira Venezuela	Araucano
P-2	Pira	Nal Tel	Confite Morocho	Araucano	Pollo
P-3	Pira	Pira Venezuela	Nal Tel	Confite Morocho	Chapalote
P-5	Nal Tel	Pira	Chirimito	Pira Venezuela	Chapalote
P-6	Confite Morocho	Nal Tel	Pira	Chapalote	Araucano
P-7	Oke	Nal Tel Blanco	Patillo Ecuatoriano	Clavito Ecuador	Pollo
P-8	Nal Tel	Pira	Confite Morocho	Chapalote	Pira Venezuela
P-9	Pira	Confite Morocho	Nal Tel	Chapalote	Pira Venezuela
P-9	Confite Morocho	Nal Tel	Pira	Pira Venezuela	Chapalote
P-10	Nal Tel	Pira	Pira Ven	Karapampa	Confite Morocho
P-11	Patillo Boliviano	Quiche§o	Oke	Bola Blanco	Altiplano
P-12	Oke	Pollo	Araucano	Aysuma	Ninuelo
P-12	Aysuma	Oke	White dent	Nal Tel Amarillo	Ninuelo
P-12	Oke	Patillo Bol	Altiplano	Quicheño	Nal Tel Blanco
P-13	Nal Tel	Karapampa	Pira	Pollo	Araucano
P-14	Oke	Aysuma	Nal Tel Amarillo	Chake-Sara	White dent
P-15	Mishca Ec	Imbricado	Patillo Ecuatoriano	Confite Puneño	Nal Tel Amarillo
P-16	Confite Puneño	Bola Blanco	Quicheño	Patillo Ecuatoriano	Canguil
P-17	Confite Morocho	Nal Tel	Pollo	Araucano	Chapalote
P-17	Araucano	Pollo	Nal Tel	Chapalote	Confite Morocho
P-18	Pira	Nal Tel	Confite Morocho	Pira Ven	Araucano
P-19	Pira	Nal Tel	Araucano	Pollo	Karapampa
P-20	Pira	Nal Tel	Confite Morocho	Pira Ven	Araucano
P-21	Pira Venezuela	Pira	Arrocillo	Chirimito	Nal Tel
P-22	Pira Venezuela	Pira	Nal Tel	Confite Morocho	Chirimito
P-23	Confite Morocho	Nal Tel	Enano Per	Polulo	Pira
P-24	Confite Morocho	Araucano	Nal Tel	Pollo	Pira

(*continued*)

Specimen	Race 1	Race 2	Race 3	Race 4	Race 5
P-25	Confite Morocho	Confite Puntiagudo	Canguil	Nal Tel	Chapalote
P-26	Confite Puntiagudo	Canguil	Enano Per	Confite Morocho	Enano Bol
P-26	Confite Morocho	Nal Tel	Enano Per	Polulo	Pira
P-27	Shajatu	Chulpi Ec	Marcame	Maranon	Perlilla Per
P-27	Chaparreño	Checchi Bol	Culli A	Chake-Sara	Marron A
P-28	Oke	Chirimito	Chapalote	Nal Tel	Nal Tel Am
P-29	Nal Tel	Pira	Chapalote	Chirimito	Pira Ven
P-30	Pira Venezuela	Arrocillo	Pororo	Enano Bolivia	Pira
P-30	Nal Tel	Chirimito	Oke	Pira	Chapalote
P-31	Nal Tel	Pira	Chapalote	Araucano	Pollo
P-32	Conico	Chutucuno Grande	Guaribero	Pisincho	Chococeño
P-33	Pira Venezuela	Pira	Nal Tel	Chirimito	Arrocillo
P-34	Nal Tel	Karapampa	Pira	Araucano	Pollo
P-35	Pira Venezuela	Nal Tel	Pira	Confite Morocho	Polulo
P-36	Pira	Nal Tel	Chapalote	Pira Ven	Chirimito
P-37	Nal Tel	Pira	Pira Ven	Confite Morocho	Chapalote
P-38	Confite Morocho	Nal Tel	Pira	Pira Venezuela	Chapalote
P-39	Nal Tel	Pira	Araucano	Pollo	Chapalote
P-40	Pira Venezuela	Pira	Nal Tel	Chirimito	Confite Morocho
P-40	Nal Tel	Chirimito	Chapalote	Yellow pop	Pira
P-41	Nal Tel	Chirimito	Pira	Chapalote	Pira Venezuela
P-42	Nal Tel	Pira	Pira Venezuela	Chirimito	Chapalote
P-42	Pira Venezuela	Pira	Nal Tel	Karapampa	Chirimito
P-42	Canguil	Confite Puntiagudo	Chapalote	Quicheño	Confite Puneño
P-43	Pira	Pira Venezuela	Nal Tel	Confite Morocho	Chapalote
P-44	Pira	Pira Venezuela	Nal Tel	Karapampa	Confite Morocho
P-45	Pira Venezuela	Pira	Nal Tel	Confite Morocho	Arrocillo
P-45	Confite Morocho	Pira	Pira Ven	Nal Tel	Polulo

P-47	Karapampa	Nal Tel	Pira	Pollo	Araucano
P-48	Chapalote	Araucano	Nal Tel	Ninuelo	Karapampa
P-49	Confite Morocho	Pira Venezuela	Nal Tel	Pira	Polulo
P-51	Ninuelo	Nal Tel Amarillo	Araguito	Aysuma	Nal Tel Blanco
P-52	Karapampa	Nal Tel	Pira	Araucano	Pollo
P-53	Choclero	Chuspillu	Capio Grande	Lenha	Paru
P-54	Paru	Choclero	Capio Grande	Chuspillu	Shajatu
P-55	Paru	Capio Grande	Choclero	Ancashino	Paro Peru
P-56	Pojoso Chico	Cholito	Guaribero	Pagaladroga	Chaparreno
P-57	Nal Tel	Chirimito	Chapalote	Enano Peru	Pira
P-58	Chapalote	Confite Morocho	Nal Tel	Canguil	Araucano
P-59	Pojoso Chico	Quicheño	Rienda	Pojoso Chico	Imbricado
P-59	Quicheño	Bola Blanco	Canguil	Patillo	Confite Puntiagudo
P-60	Güirua	Kulli	Huayleño	Aleman	Marron
P-61	Rienda	Canilla	Quicheño	Morado	Guaribero
P-61	Mochero	Patillo	Pojoso Chico	Imbricado	Guaribero
P-62	Pira Ven	Polulo	Pororo	Pira	Enano Peru
P-63	Mishca	Imbricado	Patillo	Quicheño	Nal Tel Amarillo
P-64	Confite Morocho	Pira	Nal Tel	Pira Venezuela	Karapampa
P-65	Confite Morocho	Pira	Nal Tel	Pira Venezuela	Karapampa
P-66	Araucano	Pollo	Nal Tel	Pira	Karapampa
P-67	Karapampa	Nal Tel	Pira	Pira Venezuela	Chirimito
P-67	Nal Tel	Pira	Karapampa	Pira Venezuela	Chirimito
P-68	Karapampa	Araguito	Ninuelo	Nal Tel	Pira
P-69	Nal Tel	Pira	Chapalote	Chirimito	Pira Venezuela
P-70	Pira Venezuela	Nal Tel	Pira	Confite Morocho	Karapampa
P-71	Pira	Nal Tel	Pira Venezuela	Confite Morocho	Chapalote
P-72	Pira	Nal Tel	Pira Venezuela	Karapampa	Chapalote

(*continued*)

BIBLIOGRAPHY

Aitken, M. J. 1968. "Thermoluminescent Dating in Archaeology: Introductory Review." In *Thermoluminescence of Geological Materials*, edited by D. J. McDougall. London: Academic Press.

Anderson, Edgar. 1943. "Races of Maize II: A General Survey of the Problem." *Acta Americana* 1:58–68.

———. 1946. "Maize in Mexico." *Annals of the Missouri Botanical Garden* 33:147–247.

———. 1947. "Field Studies of Guatemalan Maize." *Annals of the Missouri Botanical Garden* 34:433–67.

Anderson, Edgar, and F. D. Blanchard. 1942. "Prehistoric Maize from Cañon del Merto." *American Journal of Botany* 29:832–35.

Anderson, Edgar, and Hugh Cutler. 1942. "Races of *Zea mays* I: Their Recognition and Classification." *Annals of the Missouri Botanical Garden* 29:69–89.

Anton, Ferdinand. 1969. *Ancient Mexican Art*. New York: G. P. Putnam's Sons.

Arsandaux, H., and P. Rivet. 1921. "Contribution á l'etude de la metallurgie mexicaine." *Journal de la Société des Americanistes de Paris* n.s. 13:261–80.

Baessler, Arthur. 1902–3. *Altperuanische Kunst: Beiträege zur Archäeologie des Inca Reichs*. Vol. 3. Berlin: Verlag von A. Asher.

Bell, B. 1971. "Archaeology of Nayarit, Jalisco, and Colima." In *Handbook of Middle American Indians*, vol. 11, pt. 2, edited by Robert Wauchope. Austin: University of Texas Press.

Bennett, Wendell C. 1939. "Archaeology of the North Coast of Peru: An Account of Exploration and Excavation in the Virú and Lambayeque Valleys." *American Museum of Natural History Anthropological Papers* 37, pt. 1.

———. 1944. "The Archaeology of the Central Andes." In *Handbook of the South American Indians*, vol. 2, edited by Julian H. Steward. Bureau of American Ethnology Bulletin 143. Washington: Smithsonian Institution.

Benson, Elizabeth P. 1972a. *The Mochica*. New York: Praeger.

———, ed. 1972b. *The Cult of the Feline*. Washington, D.C.: Dumbarton Oaks Research and Library Collections.

Benz, Bruce F. 1994. "Reconstructing the Racial Phylogeny of Mexican Maize: Where Do We Stand?" In *Corn and Culture in the Prehistoric New World*, edited by Sissel Johannessen and Christine A. Hastorf. Boulder: Westview Press.

Bernal, Ignacio. 1965. "Archaeological Synthesis of Oaxaca." In *Handbook of Middle American Indians*, vol. 3, edited by Robert Wauchope. Austin: University of Texas Press.

Bird, Robert McKelvy. 1970. "Maize and Its Cultural and Natural Environment in the Sierra of Huánuco, Peru." Ph.D. dissertation, University of California, Berkeley.

———. 1978a. "Tracing the Evolution of Maize." Paper delivered at the Symposium "Origin of *Zea mays* and Its Relatives." Harvard University, Cambridge, Mass.

———. 1978b. "Archeological Maize from Peru." *Maize Genetics Cooperation Newsletter* 52:90–92.

———. 1984. "South American Maize in Central America." In *Pre-Columbian Plant Migration*, edited by Doris Stone. Cambridge: Harvard University Press.

Bird, Robert McKelvy, and Junius B. Bird. 1980. "Gallinazo Maize from the Chicama Valley, Peru." *American Antiquity* 45:325–31.

Bird, Robert McKelvy, and Major M. Goodman. 1977. "The Races of Maize V: Grouping Maize Races on the Basis of Ear Morphology." *Economic Botany* 31:471–81.

Bollaert, William. 1860. *Antiquarian, Ethnological and Other Researches in New Granada, Ecuador, Peru and Chile*. London: Trübner.

Bonavía, Duccio. 1982. *Los Gavilanes*. Lima: Editorial Ausonia–Talleres Graficos S.A.

Boos, Frank H. 1966a. *The Ceramic Sculptures of Ancient Oaxaca*. New York: A. S. Barnes.

———. 1966b. *Las urnas zapotecas de las Colecciones Frissell y Leigh del Museo Frissell de Arte Zapoteca, Mitla*. Corpus Antiquitatum Americanensium, 2. Instituto Nacional de Antropología, Mexico.

Borhegyi, Stephen. 1959. "Pre-Columbian Cultural Connections between Mesoamerica and Ecuador." *Tulane University Middle American Research Institute Records* 2, nos. 6–7.

Brieger, F. G., T. A. Gurgel, E. Paterniani, A. Blumenschein, and M. R. Alleoni. 1958. *Races of Maize in Brazil and Other Eastern South American Countries*. National Academy of Sciences Publ. 593. Washington, D.C.: National Research Council.

Brown, William L. 1960. *Races of Maize in the West Indies*. National Academy of Sciences Publ. 792. Washington, D.C.: National Research Council.

Brown, William L., and Edgar Anderson. 1947. "The Northern Flint Corns." *Annals of the Missouri Botanical Garden* 34:1–29.

Brown, William L., and Major M. Goodman. 1977. "Races of Maize." In *Corn and Corn Improvement*, edited by C. F. Sprague. Madison, Wisc.: American Society of Agronomy.

Byers, Douglas S., ed. 1967. *Prehistory of the Tehuacán Valley*, vol. 1, *Environment and Subsistence*. Austin: University of Texas Press.

Cárdenas, Martin. 1943. *Notas preliminares sobre la materia medica boliviana*. Bolivia: Publicaciones de la Universidad de Cochabamba.

Caso, Alfonso. 1928. *Las Estelas Zapotecas*. Mexico: Secretaria de Educación Publica.

———. 1938. *Exploraciones en Oaxaca, quinta y sexta temporadas 1936–37*. Instituto Panamericano de Geografía e Historia, Publ. 34.

———. 1969. *El Tesoro de Monte Albán*. Mexico: Instituto Nacional de Antropología e Historia, III.

Caso, Alfonso, and Ignacio Bernal. 1952. *Urnas de Oaxaca*. Mexico: Instituto Nacional de Antropología e Historia, II.

———. 1965. "Ceramics of Oaxaca." In *Handbook of Middle American Indians*, vol. 3, pt. 2, edited by Robert Wauchope. Austin: University of Texas Press.

Chard, Chester. 1950. "Pre-Columbian Trade between North and South America." *Kroeber Anthropological Society Papers* 1:1–27.

Cobo, Bernabé. 1890–95. *Historia del Nuevo Mundo*. Seville: E. Rasco.

Coe, Michael D. 1960. "Archaeological Linkages with North and South America at La Victoria, Guatemala." *American Anthropologist* 62:363–93.

———. 1962a. "An Olmec Design on an Early Peruvian Vessel." *American Antiquity* 27:579–80.

———. 1962b. *Mexico*. New York: Frederick A. Praeger.

———. 1966. *The Maya*. New York: Frederick A. Praeger.

Cook de Leonard, Carmen. 1971. "Minor Arts of the Classic Period in Central Mexico." In *Handbook of Middle American Indians*, vol. 10, pt. 1, edited by Robert Wauchope. Austin: University of Texas Press.

Cordy-Collins, Alana. 1994. "An Unshaggy Dog Story." *Natural History* 2:33–40.

Covarubbias, Miguel. 1957. *Indian Art of Mexico and Central America*. New York: Knopf.

Cutler, Hugh C. 1944. "Medicine Men and the Preservation of a Relict Gene in Maize." *Journal of Heredity* 35:290–94.

———. 1946. "Races of Maize in South America." *Harvard University Botanical Museum Leaflets* 12:257–91.

———. 1952. "A Preliminary Survey of Plant Remains of Tularosa Cave." *Fieldiana: Anthropology* 40:460–79.

Díaz del Castillo, Bernal. 1934. *Verdadera y Notable Relación del Descubrimiento y Conquista de la Nueva España y Guatemala*, II. Guatemala: La Tipografía Nacional.

Disselhoff, Hans Dietrich. 1967. *Daily Life in Ancient Peru*. New York: McGraw-Hill.

Doebley, John F. 1994. "Morphology, Molecules and Maize." In *Corn and Culture in the Prehistoric New World*, edited by Sissel Johannessen and Christine A. Hastorf. Boulder: Westview Press.

Donnan, Christopher B. 1965. "Moche Ceramic Technology." *Nawpa Pacha* 3:117–34.

———. 1976. *Moche Art and Iconography*. Los Angeles: UCLA Latin American Center Publications.

———. 1992. *Ceramics of Ancient Peru*. Los Angeles: University of California, Fowler Museum of Cultural History.

Drennan, Robert D. 1976. "Fabrica San José and Middle Formative Society in the Valley of Oaxaca." In *Prehistory and Human Ecology of the Valley of Oaxaca*, vol. 4, edited by Kent V. Flannery. Ann Arbor: University of Michigan Memoirs of the Museum of Anthropology No. 5.

Dunn, Mary Eubanks. 1975. "Ceramic Evidence for the Prehistoric Distribution of Maize in Mexico." *American Antiquity* 40:305–14.

———. 1977. "Races of Maize and Cultural Diffusion." Ph.D. dissertation, University of North Carolina, Chapel Hill.

———. 1978. "Suggestions for the Evaluation of Archaeological Maize Remains." *American Antiquity* 43:97–99.

———. 1979. "Ceramic Depictions of Maize: A Basis for Classification of Prehistoric Races." *American Antiquity* 44:757–74.

———. 1984. "Ceramic Evidence for the Spread of Prehistoric Races of Maize." *National Geographic Society Research Reports 1975 Projects*.

See also Eubanks, Mary W.

Edwards, Clinton R. 1960. "Sailing Rafts of Sechura: History and Problems of Origin." *Southwestern Journal of Anthropology* 16:368–91.

———. 1965a. "Aboriginal Sail in the New World." *Southwestern Journal of Anthropology* 21:351–58.

———. 1965b. *Aboriginal Watercraft on the Pacific Coast of South America*. University of California at Berkeley: Ibero-Americana 47.

———. 1969. "Possibilities of Pre-Columbian Maritime Contacts among New World Civilizations." In *Pre-Columbian Contact within Nuclear America*, edited by J. Charles Kelley and Carrol L. Riley. Carbondale: Southern Illinois University, University Museum Research Record No. 4.

Eubanks, Mary W. 1973. "A Study of the Interrelationship of the Authentication of Zapotec Urns and the Evolution of Maize." M.A. Thesis, University of North Carolina, Chapel Hill.

———. 1995. "A Cross between Two Maize Relatives: *Tripsacum dactyloides* and *Zea diploperennis* (Poaceae)." *Economic Botany* 49:172–82.

———. 1997. "Molecular Analysis of Crosses between *Tripsacum dactyloides* and *Zea diploperennis* (Poaceae)." *Theoretical and Applied Genetics* 94:707–12.

See also Dunn, Mary Eubanks.

Flannery, Kent V., and C. Earle Smith. 1983. "Monte Albán IV Foodstuffs in Guilá Naquitz Cave." In *The Cloud People: Divergent Evolution of the Zapotec and Mixtec Civilizations*, edited by Kent V. Flannery and Joyce Marcus. New York: Academic Press.

Flannery, Kent V., M. J. Kirkby, and Aubrey W. Williams. 1967. "Farming Systems and Political Growth in Ancient Oaxaca." *Science* 158:445–54.

Ford, Richard I. 1976. "Carbonized Plant Remains." In *Prehistory and Human Ecology of the Valley of Oaxaca*, vol. 4, edited by Kent V. Flannery. Ann Arbor: University of Michigan Memoirs of the Museum of Anthropology No. 5.

Friedberg, Claudine. 1958. "Contribution a l'étude ethnobotanique des tombes précolombiennes de Lauri (Pérou)." *Journal d'Agriculture Tropicale et de Botanique Appliquée* 5:397–434.

Fuhrman, Ernst. 1922. *Peru II*. Folkwang-Verlag G.M.B.H., Hagen i. W. und Darmstadt.

Galinat, Walton C. 1972. "Common Ancestry of the Primitive Races of Maize Indigenous to the Ayacucho Area in Peru." *Maize Genetics Cooperation Newsletter* 46:108–9.

———. 1998. "Maize Genetics Ten Centuries before Mendel, as Suggested by Indian Art." *Maize Genetics Cooperation Newsletter* 72:83–84.

Galinat, Walton C., and James E. Gunnerson. 1963. "Spread of Eight-Rowed Maize from the Prehistoric Southwest." *Harvard University Botanical Museum Leaflets* 20:117–60.

Gillin, John. 1945. *Moche: A Peruvian Coastal Community*. Institute of Social Anthropology, Publ. 3. Washington, D.C.: Smithsonian Institution.

Goodman, Major M. 1994. "Racial Sampling and Identification in Maize: Quantitative Genetic Variation versus Environmental Effects." In *Corn and Culture in the Prehistoric New World*, edited by Sissel Johannessen and Christine A. Hastorf. Boulder: Westview Press.

Goodman, Major M., and R. McK. Bird. 1977. "The Races of Maize IV: Tentative Grouping of 219 Latin American Races." *Economic Botany* 31:204–21.

Goodman, Major M., and E. Paterniani. 1969. "The Races of Maize III: Choices of Appropriate Characters for Racial Classification." *Economic Botany* 23:265–73.

Grant, Ulysses J., William H. Hatheway, David H. Timothy, Climaco Cassalett D, and Lewis M. Roberts. 1963. *Races of Maize in Venezuela*. National Academy of Sciences Publ. 1136, Washington, D.C.: National Research Council.

Grobman, Alexander, and Duccio Bonavía. 1978. "Pre-Ceramic Maize on the North-Central Coast of Peru." *Nature* 276:386–87.

Grobman, Alexander, Wilfredo Salhuana, and Ricardo Sevilla in collaboration with Paul C. Mangelsdorf. 1961. *Races of Maize in Peru*. National Academy of Sciences Publ. 915. Washington, D.C.: National Research Council.

Gumerman, George IV. 1994. "Corn for the Dead: The Significance of *Zea mays* in Moche Burial Offerings." In *Corn and Culture in the Prehistoric New World*, edited by Sissel Johannessen and Christine A. Hastorf. Boulder: Westview Press.

Haemig, P. D. 1979. "Secret of the Painted Jay." *Biotropica* 11:81–87.

Hatheway, William H. 1957. *Races of Maize in Cuba*. National Academy of Sciences Publ. 453. Washington, D.C.: National Research Council.

Hay, Clarence L., Samuel K. Lothrop, Harry L. Shapiro, and George C. Vaillant, eds. 1940. *The Maya and Their Neighbors*. New York: Appleton-Century.

Hébert, J. 1902. "Quelques mots sur la technique des céramistes Péruviens." *Journal de la Société des Américanistes de Paris* 4:1–7.

Helbaek, Hans. 1959. "Domestication of Food Plants in the Old World." *Science* 130:365–72.

Jijón y Caamaño. 1949. "Un vaso 'thin orange' del país Caraqui, Ecuador." *American Antiquity* 11:65–75.

Jiménez Moreno, Wigberto. 1970. "Mesoamerica Before the Toltecs." In *Ancient Oaxaca*, edited by John Paddock. Stanford: Stanford University Press.

Joyce, Thomas Athol. 1927. *Maya and Mexican Art*. London: "The Studio."

Kan, Michael. 1972. "The Feline Motif in Northern Peru." In *The Cult of the Feline*, edited by Elizabeth Benson. Washington, D.C.: Dumbarton Oaks.

Kelley, David H., and Duccio Bonavía. 1963. "New Evidence for Preceramic Maize on the Coast of Peru." *Nawpa Pacha* 1:39–41.

Kelley, J. Charles, and Carroll L. Riley. 1969. *Pre-Columbian Contact within Nuclear America*. Carbondale: Southern Illinois University, University Research Record No. 4.

Kidder, Alfred II. 1940. "South American Penetrations in Middle America." In *The Maya and Their Neighbors*, edited by Clarence L. Hay, Samuel K. Lothrop, Harry L. Shapiro, and George C. Vaillant. New York: Appleton-Century.

Kirkby, Anne V. T. 1973. "The Use of Land and Water Resources in the Past and Present Valley of Oaxaca." In *Prehistory and Human Ecology of the Valley of Oaxaca*, vol. 1, edited by Kent V. Flannery. Ann Arbor: Memoirs of the Museum of Anthropology, University of Michigan, No. 5.

Klein, Otto. 1967. *La Ceramica Mochic: Carácteres estilisticos y conceptos*. Valparaiso: La Universidad Technica "Federico Santa Maria."

Kosok, Paul. 1965. *Life, Land and Water in Ancient Peru*. New York: Long Island University Press.

Kowalewski, Stephen A., Gary M. Feinman, Laura Finsten, Richard E. Blanton, and Linda M. Nicholas. 1989. *Monte Albán's Hinterland*. Vol. 1, *Prehispanic Settlement Patterns in Tlacolula, Etla, and Ocotlan, the Valley of Oaxaca, Mexico*, pt. 2. Ann Arbor: Memoirs of the Museum of Anthropology, University of Michigan, No. 23.

Kroeber, A. L. 1925. "The Uhle Pottery Collections from Moche." *University of California Publications in American Archaeology and Ethnology* 21:191–234.

Kutscher, Gerdt. 1950. "Iconographic Studies as an Aid in the Reconstruction of Early Chimú Civilization." *Transactions of the New York Academy of Sciences*, ser. 2, 12:194–203.

Kuttruff, Carl. 1978. "Figurines and Urn Fragments from the Monte Albán Survey." In *Monte Albán: Settlement Patterns at the Ancient Zapotec Capital*, edited by Richard E. Blanton. New York: Academic Press.

Lanning, Edward P. 1963. "Olmec and Chavín: Reply to Michael D. Coe." *American Antiquity* 29:99–101.

Larco Hoyle, Rafael. 1938. *Los Mochicas*. Vols. 1 and 2. Lima: "La Cronica" y "Vaiadades," S.A.

———. 1944. "A Culture Sequence for the North Coast of Peru." In *Handbook of*

South American Indians, vol. 2, *The Andean Civilizations*, edited by Julian H. Steward. Washington, D.C.: Smithsonian Institution Bureau of American Ethnology Bulletin 143.
———. 1945. *Los Mochicas*. Buenos Aires: Sociedad Geográfica Americana.
———. 1948. *Cronología arqueológica del Norte del Perú*. Buenos Aires: Sociedad Geográfica Americana.
———. 1966. *Peru*. Translated by James Hogarth. Geneva: Nagel.
Lechtman, H. 1980. "The Central Andes: Metallurgy without Iron." In *The Coming of the Age of Iron*, edited by T. A. Wertime and J. D. Muhly. New Haven: Yale University Press.
Leigh, Howard. 1970. "The Evolution of the Zapotec Glyph C." In *Ancient Oaxaca*, edited by John Paddock. Stanford: Stanford University Press.
Lenz, Lee Wayne. 1948. "Comparative Histology of the Female Inflorescence of *Zea mays* L." *Annals of the Missouri Botanical Garden* 35:353–77.
Linares, Olga F., Payson D. Sheets, and Jane E. Rosenthal. 1975. "Prehistoric Agriculture in Tropical Highlands." *Science* 187:137–45.
Linné, S. 1925. *South American Ceramics*. Göteborg: K. Vet. O. Vitterhets-Sammhalles Handlingar vol. 29.
———. 1938. *Zapotecan Antiquities and the Paulson Collection in the Ethnographical Museum of Sweden*. Stockholm: Ethnographical Museum of Sweden New Series Publ. No. 4.
Long, A., B. F. Benz, D. J. Donahue, A. J. T. Jull, and L. J. Toolin. 1989. "First Direct AMS Dates on Early Maize from Tehuacán Mexico." *Radiocarbon* 31:1030–35.
Lothrop, Samuel K. 1932. "Aboriginal Navigation off the West Coast of South America." *Journal of the Royal Anthropological Institute* 62:229–56.
———. 1936. *Zacualpa: A Study of Ancient Quiché Artifacts*. Washington, D.C.: Carnegie Institution of Washington Publ. 472.
———. 1937. *Coclé: An Archaeological Study of Central Panama*. Cambridge: Harvard University, Memoirs of the Peabody Museum of Archaeology and Ethnology 7.
———. 1972. *The Treasures of Ancient America*. Geneva: Editions d'Art Albert Skira.
Loven, Sven. 1935. *Origins of the Tainan Culture, West Indies*. Göteborg: Elanders boktryckeri aktiebolag.
Lumbreras, Luis G. 1974. *The Peoples and Cultures of Ancient Peru*. Translated by Betty J. Meggars from the 1969 edition. Washington, D.C.: Smithsonian Institution Press.
MacNeish, Richard Stockton, Antoinette Nelken-Terner, and Angel García Cook. 1970. *Second Annual Report of the Ayacucho Archaeological-Botanical Project*. Andover, N.H.: Robert S. Peabody Foundation.
Mangelsdorf, Paul C. 1958. "Reconstructing the Ancestor of Corn." *Proceedings of the American Philosophical Society* 102:454–63.
———. 1974. *Corn: Its Origin, Evolution and Improvement*. Cambridge: Belknap Press of Harvard University Press.
Mangelsdorf, Paul C., and Robert H. Lister. 1956. "Archaeological Evidence of the

Evolution of Maize in Northwestern Mexico." *Harvard University, Botanical Museum Leaflets* 17:151–96.

Mangelsdorf, Paul C., and Gordon C. Pollard. 1975. "Archaeological Maize from Northern Chile." *Harvard University, Botanical Museum Leaflets* 24:49–64.

Mangelsdorf, Paul C., and R. G. Reeves. 1938. "The Origin of Maize." *Proceedings of the National Academy of Sciences* 24:303–12.

———. 1939. *The Origin of Indian Corn and Its Relatives*. College Station: Texas Agricultural Experiment Station Bulletin 574.

Mangelsdorf, Paul C., and Mario Sanója O. 1965. "Early Archaeological Maize from Venezuela." *Harvard University, Botanical Museum Leaflets* 21:105–12.

Mangelsdorf, Paul C., and C. Earle Smith, Jr. 1949. "New Archaeological Evidence on Evolution in Maize." *Harvard University, Botanical Museum Leaflets* 13:213–47.

Mangelsdorf, Paul C., Herbert W. Dick, and Júlian Cámara-Hernández. 1967. "Bat Cave Revisited." *Harvard University, Botanical Museum Leaflets* 22:1–31.

Mangelsdorf, Paul C., Richard S. MacNeish, and Walton C. Galinat. 1956. "Archaeological Evidence on the Diffusion and Evolution of Maize in Northeastern Mexico." *Harvard University, Botanical Museum Leaflets* 17:125–50.

———. 1967a. "Prehistoric Wild and Cultivated Maize." In *Prehistory of the Tehuacan Valley*, vol. 1, *Environment and Subsistence*, edited by Douglas S. Byers. Austin: University of Texas Press.

———. 1967b. "Prehistoric Maize, Teosinte, Tripsacum from Tamaulipas, Mexico." *Harvard University, Botanical Museum Leaflets* 22:33–63.

Marcus, Joyce. 1983. "Rethinking the Zapotec Urn." In *The Cloud People: Divergent Evolution of the Zapotec and Mixtec Civilizations*, edited by Kent V. Flannery and Joyce Marcus. New York: Academic Press.

Marcus, Joyce, and Kent V. Flannery. 1996. *Zapotec Civilization: How Urban Society Evolved in Mexico's Oaxaca Valley*. New York: Thames and Hudson.

Martinez Hernández, Juan, ed. 1929. *Diccionario de Motul Maya Español atribuido a Fray Antonio de Ciudad Real*. Mérida, Yucatán: Talleres de la Compañia Tipográfica Yucateca, S.A.

Mason, J. Alden. 1929. "Zapotec Funerary Urns from Mexico." *Museum Journal* 20:176–78.

Meggars, Betty J. 1966. *Ecuador*. London: Thames and Hudson.

Meighan, Clement W. 1969. "Cultural Similarities between Western Mexico and Andean Regions." In *Pre-Columbian Contact within Nuclear America*, edited by J. Charles Kelley and Carroll L. Riley. Carbondale: Southern Illinois University, University Research Record No. 4.

———. 1974. "Prehistory of West Mexico." *Science* 184:1254–61.

Mendizabal, Miguel O. de. 1928. *Influencia de la Sal en la distribucion geográfica de los grupos indígenas de México*. México: Museo Nacional de Arqueología, Historia y Etnografía.

Miksicek, Charles H., Robert McK. Bird, Barbara Pickersgill, Sara Donaghey, Juliette

Cartwright, and Norman Hammond. 1981. "Preclassic Lowland Maize from Cuello, Belize." *Nature* 289:56–59.

Moseley, Michael E. 1994. *The Archaeology of Peru*. London: Thames and Hudson.

Mountjoy, Joseph B. 1969. "On the Origin of West Mexican Metallurgy." In *Pre-Columbian Contact within Nuclear America*, edited by J. Charles Kelley and Carroll L. Riley. Carbondale: Southern Illinois University, University Research Record No. 4.

Munson, Patrick J. 1966. "An Annotated Bibliography of Archaeological Maize in Eastern North America." *Pennsylvania Archaeologist* 36:50–65.

Murphy, Robert Cushman. 1941. "The Earliest Spanish Advances Southward from Panama along the West Coast of South America." *Hispanic American Historical Review* 21:2–28.

Nicholson, G. Edward. 1960. "Chicha Maize Types and Chicha Manufacture in Peru." *Economic Botany* 14:290–99.

Nickerson, Norton H. 1953. "Variation in Cob Morphology among Certain Archaeological and Ethnological Races of Maize." *Annals of the Missouri Botanical Garden* 40:79–111.

Nickerson, Norton H., and Alan P. Covich. 1967. "A Collection of Maize from Darién, Panama." *Economic Botany* 20:434–40.

Oviedo y Valdés, Gonzalo Fernández de. 1851–55. *Historia general y natural de las Indias*. 4 vols. Madrid: Real Academia de la Historia.

Paddock, John. 1970. "Oaxaca in Ancient America." In *Ancient Oaxaca*, edited by John Paddock. Stanford: Stanford University Press.

Parsons, Lee A. 1962. "An Examination of Four Moche Jars from the Same Mold." *American Antiquity* 27:515–19.

Pearsall, Deborah M. 1994. "Issues in the Analysis and Interpretation of Archaeological Maize in South America." In *Corn and Culture in the Prehistoric New World*, edited by Sissel Johannessen and Christine A. Hastorf. Boulder: Westview Press.

Pickersgill, Barbara. 1972. "Cultivated Plants as Evidence for Cultural Diffusion." *American Antiquity* 37:97–104.

Porras Berrenechea, Raúl. 1937. *Las Relaciones primitivas de la conquista del Perú*. Paris.

Porter, Muriel N. 1953. *Tlatilco and the Pre-Classic Cultures of the New World*. Viking Fund Publications in Anthropology Number 19.

Rabin, Emily. 1970. "The Lambityeco Friezes." *Bulletin of Oaxacan Studies* no. 33.

Ramírez E., Ricardo, David H. Timothy, Efraín Díaz B., and U. J. Grant in collaboration with G. Edward Nicholson Calle, Edgar Anderson, and William L. Brown. 1960. *Races of Maize in Bolivia*. National Academy of Sciences Publ. 747. Washington, D.C.: National Research Council.

Rands, Robert L., and Robert E. Smith. 1965. "Pottery of the Guatemalan Highlands." In *Handbook of Middle American Indians*, vol. 2, pt. 1, edited by Robert Wauchope. Austin: University of Texas Press.

Reichel-Dolmatoff, Gerardo. 1965. *Colombia*. London: Thames and Hudson.

Reiss, Wilhelm, and Alphons Stübel. 1880–87. *The Necropolis of Ancón*. 3 vols. Berlin.

Rickards, Constantine George. 1938. "Monograph on Ornaments on Zapotec Funerary Urns." *Journal de la Société des Américanistes*, n.s. 30:147–65.

Roberts, L. M., U. J. Grant, Ricardo Ramírez E., W. H. Hatheway, and D. L. Smith in collaboration with Paul C. Mangelsdorf. 1957. *Races of Maize in Colombia*. National Academy of Sciences Publ. 510. Washington, D.C.: National Research Council.

Rowe, John Howland. 1948. "The Kingdom of Chimor." *Acta Americana* 6:26–59.

———. 1962a. "Stages and Periods in Archaeological Interpretation." *Southwestern Journal of Anthropology* 18:40–54.

———. 1962b. *Chavín Art: An Inquiry into Its Form and Meaning*. New York: Museum of Primitive Art.

Safford, William Edwin. 1917. "Food-Plants and Textiles of Ancient America." *Proceedings of the 19th International Congress of Americanists* 29:12–30.

Sánchez-González, José de Jesús. 1989. "Relationships among the Mexican Races of Maize." Ph.D. dissertation, North Carolina State University.

Sánchez-González, José de Jesús, and Major M. Goodman. 1992. "Relationships among Mexican and Some North American and South American Races of Maize." *Maydica* 37:41–51.

Saville, Marshall H. 1899. "Exploration of Zapotecan Tombs in Southern Mexico." *American Anthropologist* n.s. 1:350–62.

———. 1904. "Funeral Urns from Oaxaca." *American Museum Journal* 4:51–60.

———. 1913. "Precolumbian Decoration of the Teeth in Ecuador." *American Anthropologist* n.s. 15:377–94.

Sawyer, Allan R. 1966. *Ancient Peruvian Ceramics: The Nathan Cummings Collection*. New York: Metropolitan Museum of Art.

Schaedel, Richard P. 1951. "Mochic Murals at Pañamarca." *Archaeology* 4:145–54.

Schmidt, Max. 1929. *Kunst und Kultur von Peru*. Berlin: Imbropylaen Verlag.

Schuler-Schömig, Immina v. 1970. *Figurenfässe aus Oaxaca*. Berlin: Museum für Völkerkunde.

Shaplin, Philippa Dunham. 1975. "An Introduction to the Stylistic Study of Oaxacan Urns." M.A. Thesis, Wellesley College.

———. 1978. "Thermoluminescence and Style in the Authentication of Ceramic Sculpture from Oaxaca, Mexico." *Archaeometry* 20:47–54.

Shook, Edwin M. 1965. "Archaeological Survey of the Pacific Coast of Guatemala." In *Handbook of Middle American Indians*, vol. 2, pt. 1, edited by Robert Wauchope. Austin: University of Texas Press.

Sneath, Peter H. A., and Robert R. Sokal. 1973. *Numerical Taxonomy: The Principles and Practice of Numerical Classification*. San Francisco: W. H. Freeman.

Spinden, Herbert J. 1917. "The Origin and Distribution of Agriculture in America." *Proceedings of the 19th International Congress of Americanists* 29:269–76.

Strong, William Duncan. 1943. "Cross-Sections of the New World Prehistory." *Smithsonian Miscellaneous Collections* 104.

Strong, William Duncan, and Clifford Evans, Jr. 1952. *Cultural Stratigraphy in the Virú Valley, Northern Peru: The Formative and Florescent Epochs*. New York: Columbia Studies in Archaeology and Ethnology 4.

Taylor, R. E. 1970. "The Shaft Tombs of Western Mexico: Problems in the Interpretation of Religious Function in Nonhistoric Archaeological Context." *American Antiquity* 35:160–69.

Thompson, Donald E. 1963. "A Mold Matrix from Peru." *American Antiquity* 28:545–47.

Thompson, Raymond H., ed. 1958. *Migrations in New World Culture History*. University of Arizona Bulletin 29.

Timothy, David H., Bertulfo Peña V., Ricardo Ramírez E. in collaboration with William L. Brown and Edgar Anderson. 1961. *Races of Maize in Chile*. National Academy of Sciences Publ. 847, Washington, D.C.: National Research Council.

Timothy, David H., William H. Hatheway, Ulysses J. Grant, Manuel Torregroza C., Daniel Sarria V, and Daniel Varela A. 1963. *Races of Maize in Ecuador*. National Academy of Sciences Publ. 975. Washington, D.C.: National Research Council.

Towle, Margaret Ashley. 1952a. "Descriptions and Identifications of the Virú Plant Remains." In *Cultural Stratigraphy in the Virú Valley, Northern Peru: The Formative and Florescent Epochs*, edited by William Duncan Strong and Clifford Evans, Jr. New York: Columbia Studies in Archaeology and Ethnology 4.

———. 1952b. "Plant Remains from a Peruvian Mummy Bundle." *Harvard University, Botanical Museum Leaflets* 15:223–46.

———. 1954. "Plant Remains." In *Early Ancón and Supe Cultures*, edited by Gordon R. Willey and John M. Corbett. New York: Columbia University Studies in Archaeology and Ethnology 3.

———. 1961. *The Ethno-Botany of Pre-Columbian Peru*. Chicago: Viking Fund Publications in Anthropology No. 30.

Trimborn, Hermann. 1968. "South Central America and the Andean Civilizations." In *Pre-Columbian Religions*, edited by Walter Krickeberg, Hermann Trimborn, Werney Müller, and Otto Zerries. Translated by Stanley Davis from the 1961 edition. London: George Weidenfeld and Nicholson.

Uhle, Max. 1889. "Ausgewählte Stücke des K. Museums für Völkerkunde zur Archäologies Americas." *Museum für Völkerkunde Veroffentlichungen* I. Berlin.

———. 1913. "Die Ruinen von Moche." *Société des Américanistes de Paris* 10:95–117.

Vargas C., César. 1962. "Phytomorphic Representations of the Ancient Peruvians." *Economic Botany* 16:106–15.

Ville de Genève. 1943. *Les Collections Précolombiennes: L'Empire des Incas*. Geneva: Musée d'Ethnographie.

Von Hagen, Victor W. 1952. "The Search for the Gilded Man." *Natural History* 61:312–21ff.

Wardle, H. Newell. 1940. "Fictile Art of the Mochicas." *University Museum Bulletin* 8:15–24.

Wauchope, Robert, ed. 1955. *Seminars in Archaeology*. Memoirs of the Society for American Archaeology 11.

Weatherwax, Paul. 1954. *Indian Corn in Old America*. New York: Macmillan.

Wellhausen, E. J., Alejandro Fuentes O., Antonio Hernández Corzo, in collaboration with Paul C. Mangelsdorf. 1957. *Races of Maize in Central America*. National Academy of Sciences Publ. 511. Washington, D.C.: National Research Council.

Wellhausen, E. J., L. M. Roberts, E. Hernández X., in collaboration with Paul C. Mangelsdorf. 1952. *Races of Maize in Mexico*. Cambridge: Harvard University Bussey Institution.

West, Robert C. 1961. "Aboriginal Sea Navigation between Middle and South America." *American Anthropologist* 63:133–35.

Wilkinson, Richard G. 1975. "Techniques of Ancient Skull Surgery." *Natural History* 84:94–101.

Willey, Gordon R. 1947. "Ecuadorean Figurines and the Ceramic Mold in the New World." *American Antiquity* 13:85–86.

———. 1955. "The Interrelated Rise of the Native Cultures of Middle and South America." In *New Interpretations of Aboriginal Culture History*, edited by Betty J. Meggars and Clifford Evans, Jr. Washington,D.C.: Anthropological Society of Washington.

———. 1966. *An Introduction to American Archaeology*, vol. 1, *North and Middle America*. Englewood Cliffs, N.J.: Prentice-Hall.

———. 1971. *An Introduction to American Archaeology*, vol. 2, *South America*. Englewood Cliffs, N.J.: Prentice-Hall.

Willey, Gordon, and Philip Phillips. 1958. *Method and Theory in American Archaeology*. Chicago: University of Chicago Press.

Yacovleff, E., and F. L. Herrera. 1934. "El Mundo vegetal de los antiquos Peruanos." *Lima: Revista del Museo Nacional* 3:243–322.

Yarnell, Richard A. 1965. "Descriptive Data from the Upper Great Lakes." *Southeastern Archaeological Conference Bulletin* 3:30–31.

Zevallos Quinoñes, Jorge. 1946. "Un Diccionario Yunga." *Revista del Museo Nacional* 15:163–88.

Zimmerman, D. W. 1971. "Uranium Distributions in Archeologic Ceramics: Dating of Radioactive Inclusions." *Science* 174:818–19.

INDEX